Fatty Alcohols
Anthropogenic and Natural Occurrence in the Environment

Fatty Alcohols
Anthropogenic and Natural Occurrence in the Environment

Stephen M Mudge
School of Ocean Sciences, Bangor University, UK

Scott E Belanger
Central Product Safety, Procter & Gamble, USA

Allen M Nielsen
SASOL-North America, USA

RSCPublishing

ISBN: 978-0-85404-152-7

A catalogue record for this book is available from the British Library

Published by The Royal Society of Chemistry,
Thomas Graham House, Science Park, Milton Road,
Cambridge CB4 0WF, UK

Registered Charity Number 207890

For further information see our web site at www.rsc.org

Preface

Fatty alcohols are widespread in the environment coming from a range of natural sources including bacteria, plants and animals. These compounds are also manufactured by industry from natural fatty acid sources or from petroleum-derived carbon. This book presents their environmental occurrence, fate and behaviour. The principal focus of past research has been on their natural production, which occurs in all living organisms from bacteria to humans, and the profiles and concentrations of these compounds in water, soils and sediments. Their relatively non-polar nature means they are principally associated with solid phases in aquatic systems (*e.g.* sediments) rather than dissolved in water. The major biological synthetic pathway is from the reduction of fatty acids, through aldehyde intermediates, to fatty alcohols and in many organisms to esters with fatty acids to form waxes. These waxes are used by organisms for a variety of purposes, from the prevention of desiccation in the terrestrial environment to energy reserves in the marine environment. They are ubiquitous and occur in most environments around the world, including the deep ocean and in sediment cores.

Due to the nature of the synthetic pathway using acetyl-CoA, most fatty alcohols are of an even chain length. Terrestrial plants utilise fatty alcohols as a waxy coating, dominated by long chain moieties with chain lengths from C_{22} to C_{32}. In contrast, marine organisms synthesise smaller compounds with peak chain lengths of C_{14} to C_{16}. Bacteria also produce fatty alcohols but these can also be odd chain lengths and contain branches. This aspect of their occurrence enables them to be used as biomarkers for organic matter sources. As well as their natural production and occurrence, fatty alcohols are also utilised in detergent formulations, principally as sulfates or polyethoxylates.

The analytical method preferred by contemporary environmental scientists used to measure the concentration of the ethoxylates involves direct derivatisation with a pyridinium complex and quantification *via* LC-MS (liquid chromatography–mass spectrometry). This technique will detect free fatty

---

Fatty Alcohols: Anthropogenic and Natural Occurrence in the Environment
By Stephen M Mudge, Scott E Belanger, and Allen M Nielsen
© Copyright 2008 ERASM (the joint surfactant environmental research platform of AISE and CESIO) and SDA
Published by the Royal Society of Chemistry, www.rsc.org

alcohols as well as the ethoxylates, but will not detect any of the bound alcohols such as the waxes. To detect this latter group, a saponification step is required. This second method, frequently using GC-MS (gas chromatography–mass spectrometry), in combination with the LC method will detect all of the ethoxylates and may be considered as giving a good measure of the total fatty alcohols present in a system.

The concentration of individual fatty alcohols in the environment ranges from low values in old deep cores and the open ocean floor (undetectable to $12\,\text{ng}\,\text{g}^{-1}$ dry weight (DW) for C_{16}) to high values near natural sources and especially in suspended particulate matter ($2.7 \times 10^6\,\text{ng}\,\text{g}^{-1}$ DW for C_{16}); this is almost a factor of 10^6 difference in their concentrations. The short chain compounds are more readily degradable than the longer chain compounds and, in many cases, are removed first as a preferred food source for bacteria. The longer chain compounds may also degrade to short chain compounds with time but, in general, the $>C_{20}$ class of alcohols degrades more slowly in sediments and soils than the $<C_{20}$ class.

The different compound profiles for each source has made them suitable as biomarkers and the use of multivariate statistical methods can clearly distinguish compounds from each potential source as well as sites. Principal component analysis (PCA) is particularly useful in this regard. Signature analysis using partial least squares (PLS) analysis has been used successfully to discriminate between samples that are impacted by marine *versus* terrestrial sources. However, due to the commonality of fatty alcohol detergent formulations and the natural environmental alcohols, source partitioning on the basis of compounds alone is not as successful. When ascribing proportions to such sources, a different approach such as stable isotopes may be more appropriate.

Based on the toxicity and ecotoxicity testing of fatty alcohols, they are relatively benign in the environment due to low environmental exposures. Free fatty alcohols have been shown to undergo very rapid and complete biodegradation. Fatty alcohols lack effects on genotoxicity and reproductive and developmental toxicity, and carcinogenicity. Health studies for oral, ingestion and inhalation exposure have all shown good margins of safety for human health and are, therefore, unlikely to lead to effects on the aquatic ecosystem.

Contents

Fatty Alcohols: Anthropogenic and Natural Occurrence in the Environment
By Stephen M Mudge, Scott E Belanger, and Allen M Nielsen
© Copyright 2008 ERASM (the joint surfactant environmental research platform of AISE and CESIO) and SDA
Published by the Royal Society of Chemistry, www.rsc.org

Chapter 6 Analytical Methods

Chapter 7 Environmental Concentrations

Acknowledgements

The authors are deeply appreciative of discussions and research collaborations with many individuals over the past several years that have culminated in the contents of this book. We especially appreciate the interactions with Hans Sanderson, Kathy Stanton, Rich Sedlak and Paul DeLeo (The Soap and Detergent Association (U.S.)), Martin Selby and Charles Eadsforth (Shell Global Solutions), Thorsten Wind (Henkel), Dennis Walker (Procter and Gamble Company and chair of the APAG Alcohols Technical Committee) and Andreas Willing (Cognis). Funding and collaborations across a variety of professional and industry associations are acknowledged including the US SDA, ERASM (European Risk Assessments and Surfactants Management; the joint surfactant environmental research platform for AISE (Association Internationale de la Savonnerie, de la Détergence et des Produits d'Entretien) and CESIO (Comite Europeen des Agents de Surface et de leurs Intermediaires Organiques)), APAG (Association Européenne des Producteurs d'Acides Gras) and the ICCA (International Council of Chemical Associations) Aliphatic Alcohol Consortium.

We would also like to thank the many researchers from several universities and institutes who have analysed these compounds over the years.

About the Authors

Stephen Mudge is a senior lecturer at Bangor University where he has studied the occurrence and use of fatty alcohols as lipid biomarkers in the environment for several years. This has principally been for the tracking of organic matter from several sources in the marine environment. He developed and ran the world's first undergraduate degree programme in environmental forensics and has acted as an expert witness is cases where it has been important to identify where potential contaminants have arisen.

Scott Belanger is a research fellow in The Procter & Gamble Company corporate environmental safety organisation. His research spans a wide range of topics including understanding the effects of consumer product chemicals in the environment at the levels of the organism to the ecosystem. He has assisted in several efforts to assess the environmental risk of alcohols and alcohol-derived surfactants in recent years frequently working with trade associations, academic partnerships and the regulatory community on these affairs.

Allen Nielsen is a recently retired microbiologist from the Research and Development Department of Sasol North America, Inc. His main focus during his 31-year career has been the environmental safety of petrochemical-derived surfactants which are used in consumer and industrial applications. In recent years he was focused on the environmental safety of alcohols and alcohol-derived surfactants.

CHAPTER 1
Definitions

This chapter aims to introduce the family of compounds, how they are referred to, the likely structures that will be found and their chemistry from an environmental point of view.

1.1 Names and Structures

Fatty alcohol is a generic term for a range of aliphatic hydrocarbons containing a hydroxyl group, usually in the terminal or *n*-position. The accepted definition of fatty alcohols states that they are naturally derived from plant or animal oils and fats and used in the pharmaceutical, detergent or plastics industries (*e.g.* Dorland's Illustrated Medical Dictionary). However, it is possible to find the hydroxyl (–OH) group in other positions within the aliphatic chain, but these secondary or tertiary alcohols are not discussed to any great extent in this book.

The generic structure of fatty alcohols or *n*-alkanols can be seen in Figure 1.1 and specific examples in Figure 1.2. The value of the *n*-component is variable and is discussed below.

The range of chain lengths for these *n*-alkanols can be from 8 to values in excess of 32 carbons. With such a wide range of chain lengths, the chemical properties and consequently the environmental behaviour vary considerably. As well as these straight chain moieties, a range of branched chain compounds are also naturally produced by micro-organisms in the environment. The major positions for the methyl branches are on the carbons at the opposite end of the molecule to the terminal –OH. If the methyl branch is one position in from the end of the molecule (ω-1), it is termed an *iso* fatty alcohol; if it is two in from the end (ω-2), it is called an *anteiso* fatty alcohol. Examples of these branches can be seen in Figure 1.2.

Most fatty alcohols are saturated in that they have no double bonds present in their structure. However, there are a limited number of mono-unsaturated compounds that can be found in nature. The two most common compounds are phytol (3,7,11,15-tetramethyl-2-hexadecen-1-ol), an isoprene[1] derived from

Fatty Alcohols: Anthropogenic and Natural Occurrence in the Environment
By Stephen M Mudge, Scott E Belanger, and Allen M Nielsen
© Copyright 2008 ERASM (the joint surfactant environmental research platform of AISE and CESIO) and SDA
Published by the Royal Society of Chemistry, www.rsc.org

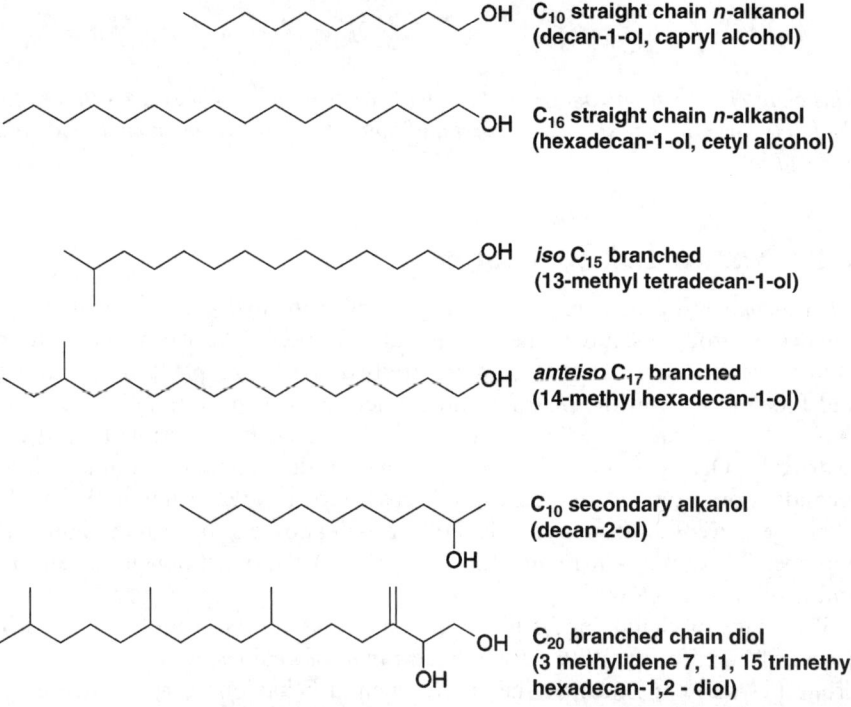

Figure 1.1 Generic structure of a fatty alcohol. The total number of carbons needs to be greater than 8–10 to be a "fatty" alcohol; shorter chain compounds have an appreciable water solubility and are generally just called alcohols.

C_{10} straight chain *n*-alkanol
(decan-1-ol, capryl alcohol)

C_{16} straight chain *n*-alkanol
(hexadecan-1-ol, cetyl alcohol)

iso C_{15} branched
(13-methyl tetradecan-1-ol)

anteiso C_{17} branched
(14-methyl hexadecan-1-ol)

C_{10} secondary alkanol
(decan-2-ol)

C_{20} branched chain diol
(3 methylidene 7, 11, 15 trimethyl hexadecan-1,2 - diol)

Figure 1.2 Example fatty alcohol structures. The majority found in nature are of the straight chain type with smaller amounts of the branched chain compounds also being present. Secondary alcohols and diols are relatively uncommon.

the side chain of chlorophyll (Figure 1.3), and a straight chain C_{20} alcohol with a double bond in the ω9 position counted from the terminal carbon (eicos-11-en-1-ol; Figure 1.3).[2]

There have been occasional reports of polyunsaturated fatty alcohols, but these are relatively rare[3] and are confined to di-unsaturates such as octadeca-dienol, 18:2. There is a group of isoprenoid lipids which may be found in bacteria and are essentially repeating isoprene subunits strung together and terminated by a hydroxyl group.[4] These compounds are also uncommon in environmental analyses and are not reported to any great extent.

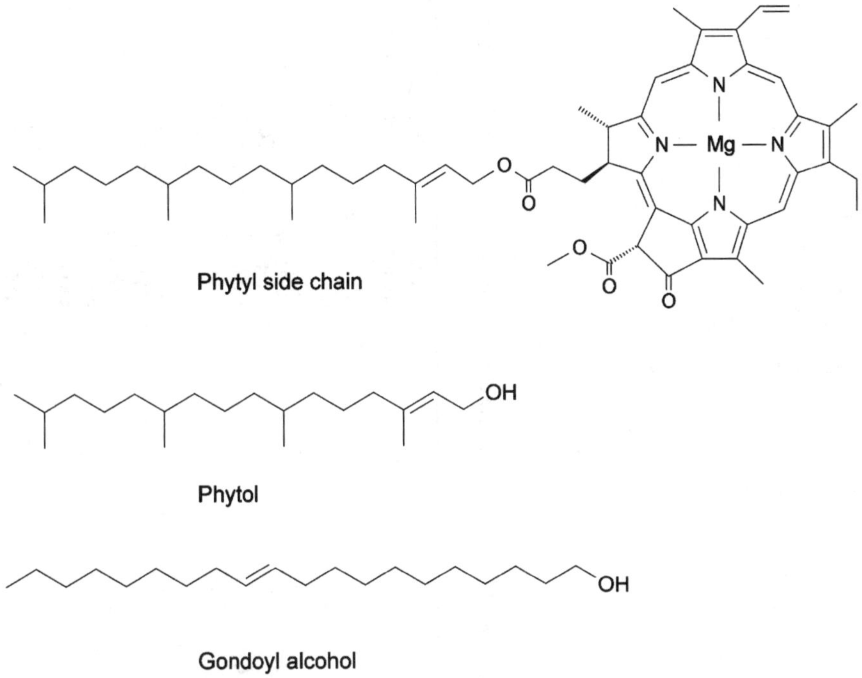

Phytyl side chain

Phytol

Gondoyl alcohol

Figure 1.3 Chlorophyll-a molecule with the phytyl side chain labelled. Cleavage of this chain at the COO– group produces free phytol in the environment. Eicos-11-en-1-ol or 20:1 fatty alcohol is one of the most frequently observed straight chain mono-unsaturated alcohols in the environment.

Fatty alcohols together with many other groups of compounds have both systematic and trivial or common names. The common name is based on the length of the alkyl chain and the root is common between aliphatic hydrocarbons and fatty acids. These common names together with the systematic name and carbon number are shown in Table 1.1.

1.2 Physicochemical Properties

1.2.1 Solubility *Versus* Chain Length

One of the key factors in determining the environmental behaviour of any compound is its water solubility; this will determine the partitioning between solid and solution phases. Compounds with low water solubility will be preferentially adsorbed to particulate matter, either settled or suspended in water. These compounds will also partition into the lipid phase of organisms and would have higher bioconcentration factors if not offset by metabolism. The available physicochemical properties for the fatty alcohol series from C_4 to C_{30}

Table 1.1 Names and key properties of fatty alcohols from C_4 to C_{30}. The horizontal line indicates the arbitrary division between alcohols and fatty alcohols.

Systematic name	Common name (excluding "alcohol" part)	Carbon number	CAS registry number	d^{20} ($g\,cm^{-3}$)	Melting point (°C)	Boiling point[a] (°C)	Water solubility[b] (at 25°C)	Predicted water solubility[c] ($mg\,l^{-1}$)
Butanol	Butyl	4	71-36-3	0.810	−90	117	$91\,ml\,l^{-1}$	
Pentanol	Amyl	5	71-41-0	0.815	−79	137.5	$27\,g\,l^{-1}$	
Hexanol	Caproyl	6	111-27-3	0.815 at 25°C	−51.6	157	$5.9\,g\,l^{-1}$	7671.3
Heptanol	Oenantyl	7	111-70-6	0.819 at 25°C	−34.6	175.8	$1.6\,g\,l^{-1}$	1743.6
Octanol	Caprylic	8	111-87-5	0.827	−16	194	$0.5\,g\,l^{-1}$	400.7
Nonanol	Pelorgonyl	9	143-08-8	0.828	−5	215	$0.1\,g\,l^{-1}$	94.2
Decanol	Capryl	10	112-30-1	0.830	6.4	232.9	$0.04\,g\,l^{-1}$	22.9
Undecanol		11	112-42-5	0.832	15	244	$0.008\,g\,l^{-1\,d}$	5.8
Dodecanol	Lauryl	12	112-53-8	0.831 at 24°C	24	259	BML	1.6
Tridecanol		13	112-70-9	0.915	30	278	BML	0.451
Tetradecanol	Myristyl	14	112-72-1	0.824	38	289	BML	0.141
Pentadecanol		15	629-76-5	0.893	42		BML	0.0490

Hexadecanol	Cetyl	16	36653-82-4	0.811	49	344	0.0190
Heptadecanol	Margaryl	17	1454-85-9	0.885	53		0.00842
Octadecanol	Stearyl	18	112-92-5	0.811	59	360	0.00433
Nonadecanol		19	1454-84-8	0.882	62		BML
Eicosanol	Arachidyl	20	629-96-9	0.88	65		BML
Henicosanol		21	15594-90-8		68.5		BML
Docosanol	Behenyl	22	661-19-8	0.8	70		BML
Tricosanol		23	3133-01-5		72		BML
Tetracosanol	Lignoceryl	24	506-51-4		72		BML
Pentacosanol		25	26040-98-2		75		BML
Hexacosanol	Cerotyl	26	506-52-5		73		BML
Heptacosanol	Carboceryl	27	2004-39-9		80		BML
Octacosanol	Montanyl	28	557-61-9		81		BML
Nonacosanol		29	6624-76-6		83.5		BML
Tricontanol	Melissyl	30	593-50-0		87		BML

[a]The boiling point values quoted are at atmospheric pressure.

[b]BML, below measurement limits.

[c]As the water solubility becomes harder to measure, the predicted water solubility is frequently used instead. The values presented here are taken from Fisk *et al.*[5] and use the ECOSAR model.

[d]At 20 °C.

are summarised in Table 1.1. These data are drawn from many sources, but principally from the online Beilstein Chemical Database (Elsevier MDL). The density and melting points in the summary data (Table 1.1) have a degree of uncertainty about them as some compounds, especially the longer chain and odd carbon number moieties, are less well studied. Density data are not available for all compounds.

The short chain compounds (up to C_9) have appreciable water solubility (Table 1.1) and would not be classified as "fatty" alcohols as the free compounds are more likely to have a substantial amount in solution rather than in the solid phase (abiotic or biotic). Compounds with a chain length greater than 10 carbons are essentially insoluble in water and will partition on to the solid phase in the environment.

1.2.2 Partitioning (K_{ow}) and Sediment Associations

It is usual to measure the water solubility and related factors such as bio-concentration factors (BCFs) through the octanol–water partition coefficient (K_{ow}) or its logarithm (log K_{ow}). There is relatively little information published for *measured* K_{ow} values for fatty alcohols, although there are some data estimated from HPLC retention times.[6] Difficulties arise in the measurement of these coefficients due to the hydrophobic–hydrophilic nature of the different parts of the molecule (Figure 1.4). The hydroxyl group gives that end of the molecule a degree of water solubility while the alkyl carbon chain is hydrophobic. Therefore, these compounds sit at the interface of octanol and water in the experimental situation.

The log K_{ow} values for compounds (Table 1.2; shown graphically in Figure 1.5) with a chain length greater than C_9 are above 4, which is indicative of materials that will be preferentially absorbed to particulate matter. In most environmental situations, this means the compounds will be associated mainly with particles such as settled and suspended sediments. The nature of these particulate materials is that they will settle out to the benthos at some stage and will be transferred

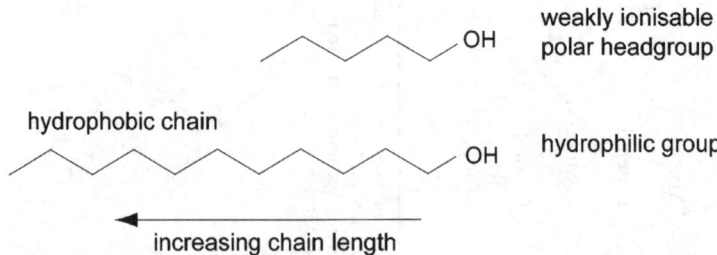

Figure 1.4 The –OH group is weakly ionisable to form –O$^-$ and H$^+$ and as such will "dissolve" in water. However, with increasing alkyl chain length, the effect of this is diminished and the compound has lower water solubility. This property allows the molecule to be used as a detergent, one of the principal anthropogenic functions of fatty alcohols.

Table 1.2 Octanol–water partition coefficients.

Fatty alcohol	Carbon number	Log K_{ow}
Butanol	4	0.785^{a}
		0.84^{c}
Pentanol	5	1.53^{a}
Hexanol	6	2.03^{a}
		1.84^{c}
Heptanol	7	2.57^{a}
		2.34^{c}
Nonanol	9	3.31^{b}
		3.77^{a}
Dodecanol	12	5.36^{b}
Tetradecanol	14	6.03^{b}
Hexadecanol	16	6.65^{b}
Octadecanol	18	7.19^{b}
Eicosanol	20	7.75^{b}

[a]From Tewari *et al.*[7]
[b]From Burkhard *et al.*[6]
[c]From Hansch *et al.*[8]

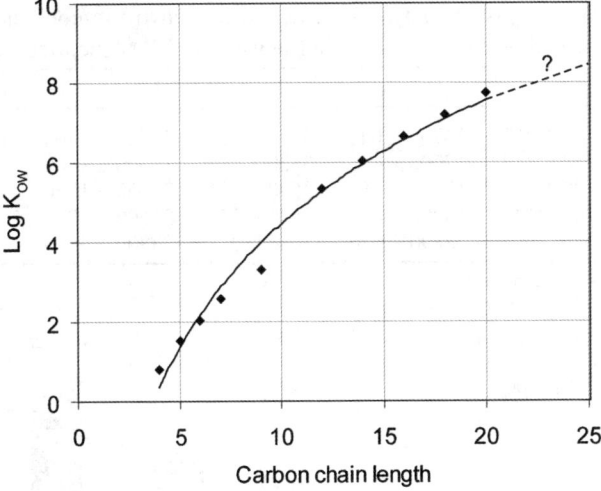

Figure 1.5 Effect of carbon chain length in fatty alcohols on log K_{ow}.

to the geosphere. This partitioning between the solution phase for short chain compounds and solid phase for long chain compounds may lead to the separation of mixtures such that short chain moieties will remain in solution while longer chain moieties may settle out. There will also be different degradation steps possible as materials in the solid phase may enter anaerobic environments in sediments; this may lead to preservation of some materials and differential products of degradation.

The association of fatty alcohols with suspended matter will be of importance in sewage treatment plants as incoming materials may be removed from the system by partitioning into the solid phase which subsequently settles out. Experiments using radiolabelled alcohols with activated sewage sludge[9] measured the time dependent partition coefficients (K_d) for a range of alcohols typically used in detergent formulations (Table 1.3). The mean K_d values can be seen in Figure 1.6; the data are presented on a logarithmic axis and a linear relationship can be seen in this figure. These values are relatively high implying that, in such a system, free fatty alcohols will be actively scavenged by the particulate phase and may be removed with the sludge or associated with suspended solids in wastewater. Thus, alcohols may leave sewage treatment plants either bound or unbound (free). Fisk *et al.*[5] using the wastewater treatment model SIMPLETREAT (a module in the European environmental distribution model EUSES) demonstrated that as carbon number increases, the fraction of fatty alcohol that is degraded by microbes in wastewater declines as the amount sorbed increases. Federle and Itrich[10] postulated that eventually at chain lengths of 16 and greater, the equilibrium desorption controls the biodegradation rate.

Table 1.3 The average (± 1 standard deviation) partition coefficients (K_d) for fatty alcohols with activated sewage sludge suspended in river water.

Time (h)	C_{12}	C_{14}	C_{15}	C_{16}	C_{18}
1	4100 ± 267	14700 ± 645	4070 ± 387	34100 ± 1700	107000 ± 6330
5	3410 ± 119	12700 ± 675	3820 ± 183	33300 ± 1600	90300 ± 3070
16	3320 ± 276	10500 ± 167	3590 ± 104	28600 ± 1720	89900 ± 1980
30	3100 ± 143	10200 ± 670	3480 ± 77	27600 ± 1930	82400 ± 2970
72	3000 ± 78	8490 ± 916	3080 ± 271	23800 ± 3160	78700 ± 5350

Data from van Compernolle *et al.*[9]

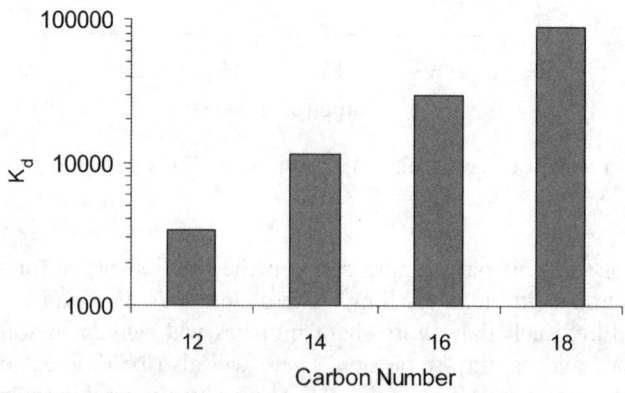

Figure 1.6 Mean K_d for the even carbon numbered fatty alcohols (data from Table 1.3). Note that the *y*-axis has a logarithmic scale.

Summary

- Fatty alcohols found in the environment are principally linear with a terminal hydroxyl group.
- As the alkyl chain becomes longer, the water solubility decreases leading to a wide range of octanol–water partition coefficients.
- Compounds with a chain length greater than C_{10} are more likely to be associated with the solid phase in the aquatic environment and become coupled with sediments and soils in the environment.
- These lower water solubility compounds will tend to bioaccumulate more than their short chain moieties, although all may be metabolised by bacteria.

CHAPTER 2
Biological Synthesis

The biochemical mechanisms that lead to fatty alcohol formation highlight the differences between bacteria that can produce odd chain and branched compounds, while most other biota produce even chain compounds.

The synthesis of fatty alcohols by living organisms is intimately linked to the production of fatty acids in most cases. In order to understand the types of fatty alcohols present in the environment, it is necessary to appreciate the biochemical synthetic pathways that lead to their formation in the first place.

The formation of fatty acids can progress through two major pathways. Animals, fungi and some mycobacteria use the Type I synthetic pathway. In this pathway, the synthesis takes place within a large single protein unit and has a single product in the form of a C_{16} unsaturated fatty acid (palmitic acid). This system has genetic coding in one location. In contrast, plants and most bacteria use a series of small discrete proteins to catalyse individual steps within the synthesis, which is termed Type II fatty acid synthesis.[11] These proteins are genetically encoded in several different locations. Yeasts are intermediate between these two extremes, where the synthesis activities take place in two separate polypeptides.[12]

2.1 Type I Fatty Acid Synthesis

Type I fatty acid synthesis (FAS) occurs in animals. As well as having this initial style of fatty acid synthesis, there are a series of subsequent reactions which lead to the elongation of the primary fatty acid (hexadecanoic acid, C_{16}) to higher carbon numbers and desaturation mechanisms leading to mono-unsaturated products. However, animals are unable to manufacture all the

Fatty Alcohols: Anthropogenic and Natural Occurrence in the Environment
By Stephen M Mudge, Scott E Belanger, and Allen M Nielsen
© Copyright 2008 ERASM (the joint surfactant environmental research platform of AISE and CESIO) and SDA
Published by the Royal Society of Chemistry, www.rsc.org

fatty acids they require and these must be obtained from plants in the diet (*e.g.* ω3 essential fatty acids).

The synthesis of fatty acids in this system occurs on a single large complex comprised of seven polypeptides. This complex acts as the focus for a series of reactions building the fatty acids up from an acetyl-CoA starter with malonyl-CoA subunits. The key components in the system can be seen in Figure 2.1. The complex performs four steps each time two carbons are added to the chain: initially CO_2 is removed from the malonyl-CoA in a condensation reaction joining the two molecules together. NAD(P)H is used in a reduction step converting the C=O group to C–OH. This is dehydrated (removal of H_2O) making a mid-chain double bond that undergoes a final reduction step with more NAD(P)H leading to a saturated alky chain.

The net effect of this series of four sub-reactions can be seen in Figure 2.2 as the product of the first step. The process is repeated until a 16-carbon chain has been created. The completed fatty acid is then cleaved from the FAS complex and is available for further reactions. This process explains why the most common fatty acid (and frequently fatty alcohol) found in environmental systems is comprised of 16 carbons. In some cases, an extra cycle occurs and a C_{18} fatty acid is formed instead.

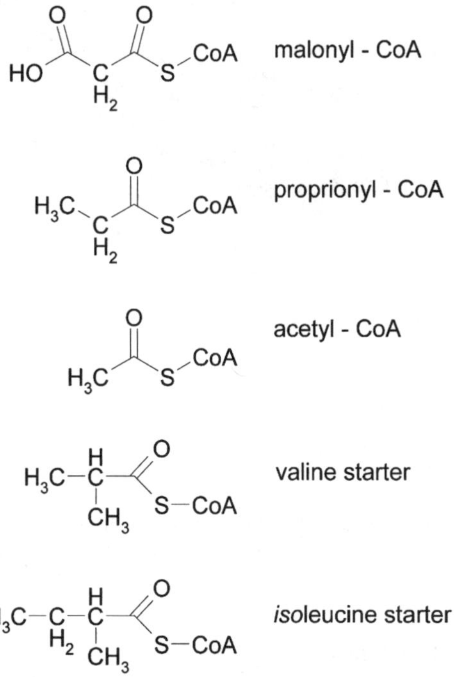

Figure 2.1   Key compounds in fatty acid synthesis. In general, plants and animals principally use acetyl-CoA as the starter, while bacteria, plants and animals may sometimes use the others as well.

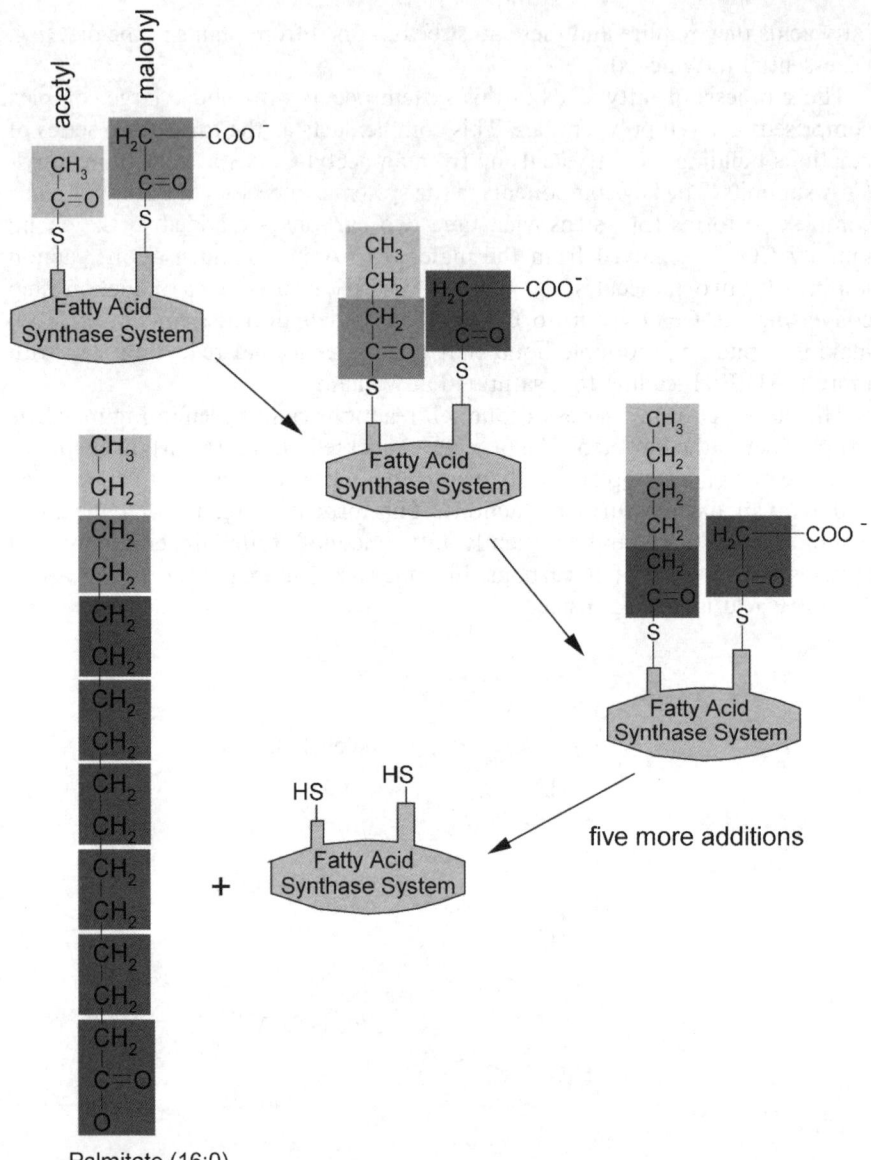

Palmitate (16:0)

Figure 2.2 Process of palmitate (C$_{16}$ fatty acid) synthesis through sequential addition of C$_2$ units from malonyl-CoA to an initial acetyl-CoA.

2.1.1 Unsaturated Chains

In animals, fatty acyl-CoA desaturase catalyses the removal of two hydrogen atoms from the bond between C$_9$ and C$_{10}$ in either palmitic or stearic acid to provide the Δ^9 *cis* double bond in palmitoleic or oleic acid (Figure 2.3).

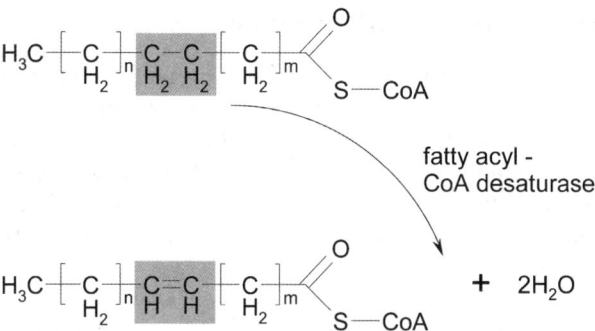

Figure 2.3 Desaturation of the acyl chain. Animals can only desaturate bonds in the Δ^9 position and closer to the carboxylic acid group. Plants are able to desaturate bonds closer to the ω end of the molecule.

Table 2.1 Fatty acid composition (as percentages) of several potential oil and fat sources. The saturated/unsaturated fatty acid ratio is also included. The sum of each row may not be 100% as minor fatty acids are also present.

Oil/fat	Saturated/ unsaturated ratio	10:0	12:0	14:0	16:0	18:0	18:1ω9	18:2ω6	18:3ω3
Beef tallow	0.9	–	–	3	24	19	43	3	1
Butterfat (cow)	0.5	3	3	11	27	12	29	2	1
Canola (rapeseed)	15.7	–	–	–	4	2	62	22	10
Cod liver oil	2.9	–	–	8	17	–	22	5	–
Coconut oil	0.1	6	47	18	9	3	6	2	–
Corn oil	6.7	–	–	–	11	2	28	58	1
Cottonseed oil	2.8	–	–	1	22	3	19	54	1
Flaxseed oil	9.0	–	–	–	3	7	21	16	53
Lard (pork fat)	1.2	–	–	2	26	14	44	10	–
Olive oil	4.6	–	–	–	13	3	71	10	1
Palm oil	1.0	–	–	1	45	4	40	10	–
Palm kernel oil	0.2	4	48	16	8	3	15	2	–
Peanut oil	4.0	–	–	–	11	2	48	32	–
Safflower oil	10.1	–	–	–	7	2	13	78	–
Sesame oil	6.6	–	–	–	9	4	41	45	–
Soybean oil	5.7	–	–	–	11	4	24	54	7
Sunflower oil	7.3	–	–	–	7	5	19	68	1

Table 2.1 shows the fatty acid content of several major oil and fat sources. In general, the animal sources have low quantities of unsaturated fatty acids while the plant sources have large amounts of polyunsaturated compounds. For example, for beef tallow the ratio of the saturated compounds to the unsaturated compounds is 0.9 compared to 15.7 for Canola or rapeseed oil. The exceptions to this are the oils derived from coconuts or palms which have large quantities of short chain unsaturated fatty acids making them amenable as a feedstock in some industrial processes (see Chapter 4).

2.2   Type II Fatty Acid Synthesis

Type II FAS in bacteria and plants occurs in a similar fashion to Type I, but the seven different polypeptides are independent of one another. The reactions are similar to those discussed above, but the products may then undergo a wider range of elongation and desaturation reactions. In the case of some plants (*e.g.* coconuts and palms; Figure 2.4), the fatty acid may be cleaved before it reaches 16 carbons and up to 90% of the oil from these plants may have fatty acids between C_8 and C_{14}.[12]

2.2.1   Unsaturated Compounds

Unlike most animals, plants can introduce double bonds into fatty acids at locations other than the Δ^9 position. They have enzymes that act on the Δ^{12} and Δ^{15} positions of oleic acid ($18:1\omega9$) but only when it is part of a phospholipid or phosphatidylcholine. This specificity may explain why very few poly-unsaturated fatty alcohols are found.

Plants frequently contain fatty acids with two or more double bonds within the molecule (Table 2.1). For example, the principal fatty acid within linseed oil is linolenic acid or $18:3\omega3$, an 18-carbon straight chain molecule with three double bonds, the first of which is in position three from the ω end of the molecule (Δ^{15}). Animals cannot generally make these polyunsatura-ted compounds and must obtain them from their diet, hence their being refer-red to as essential fatty acids. Once in animals, however, they may be elongated to form a range of other biochemically active compounds such as prosta-glandins.[12]

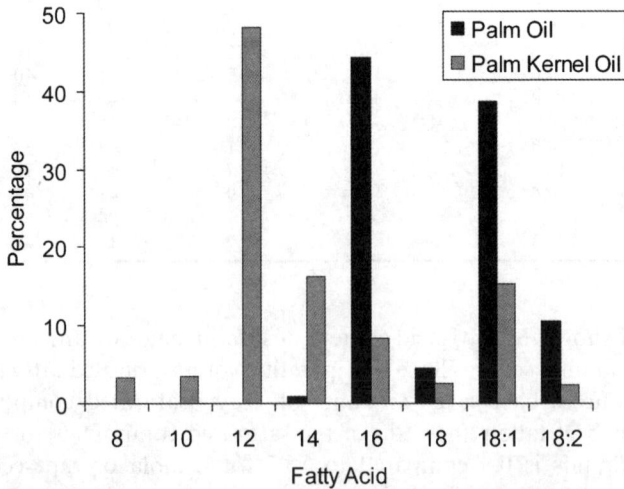

Figure 2.4   Fatty acid chain length profile of palm oil and palm kernel oil; the former has a substantial quantity of fatty acids with chain length less than 16.

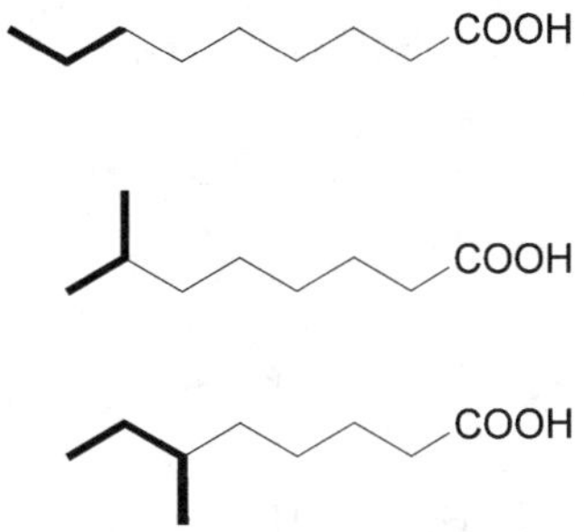

Figure 2.5 Orientation of the precursor leading to the formation of branched odd chain length fatty acids.

2.2.2   Branched Chains

Bacteria make branched chain fatty acids and alcohols;[13] the orientation of the carbons in the starter complex during the initial stages of FAS determines the final structure. There are three possible orientations which yield either a straight chained odd carbon numbered compound or two branched compounds with the methyl group in the *iso* or *anteiso* position (Figures 1.2 and 2.5).

It is also possible to start the fatty acid synthesis with an amino acid.[14] The structure of the most appropriate molecules, valine and *iso*leucine, are shown in Figure 2.1. When valine is used, an *iso*-branched product of the FAS is formed while *iso*leucine yields *anteiso*-branched products. These compounds are the principal fatty acids in Gram-positive bacteria.[15] The chain elongation process is the same as other higher organisms (*e.g.* Figure 2.2) but the starter compounds are different. Mid-chain branches are derived from other pathways where typically a methyl subunit is added across the double bond of an unsaturated compound such as oleic acid.[15]

2.3   Fatty Acid Degradation

The C_{16} fatty acid produced by the Type I and Type II FAS pathways may undergo chain shortening (Figure 2.6) as well as elongation and desaturation. This is particularly important with regard to the formation of some fatty alcohols (see below) that require an appropriate fatty acid to start with. There are several enzyme systems involved in the process belonging to the acyl-CoA dehydrogenase family: those fatty acids with carbon chains between C_{12} and C_{18} use a long chain acyl-CoA dehydrogenase, a medium chain one operates on C_4 to C_{14} acids while a short chain one acts on C_4 to C_6 only.[16]

2.4   Fatty Acyl-CoA Reductase (FAR)

Fatty alcohols have several uses within an organism; they are principally associated with waxes and storage lipids although ether lipids also contain

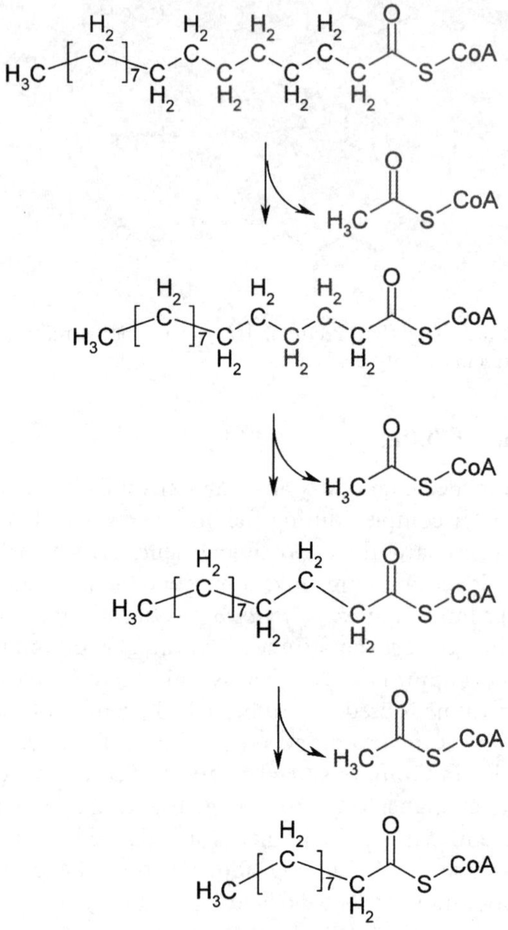

Figure 2.6   Sequence by which fatty acids (as a CoA) are shortened by fatty acyl-CoA dehydrogenase. Two carbon subunits are sequentially removed.

alcohols.[17] Waxes are abundant neutral lipids that coat the surfaces of plants, insects and mammals. They are composed of long chain alcohols linked *via* an ester bridge to fatty acids and have the chemical property of being solid at room temperature and liquid at higher temperatures. Waxes have several essential biological roles including preventing water loss, abrasion and infection.[18]

According to Cheng and Russell,[19] who studied the synthesis of wax in mammals, two catalytic steps are required to produce a wax monoester (Figure 2.7). These include a reduction step of a fatty acid to a fatty alcohol and subsequently the transesterification of the fatty alcohol to a fatty acid. The first step is catalysed by the enzyme fatty acyl-CoA reductase (FAR) which uses the reducing equivalents of NAD(P)H to convert a fatty acyl-CoA into a fatty alcohol and Co-ASH.

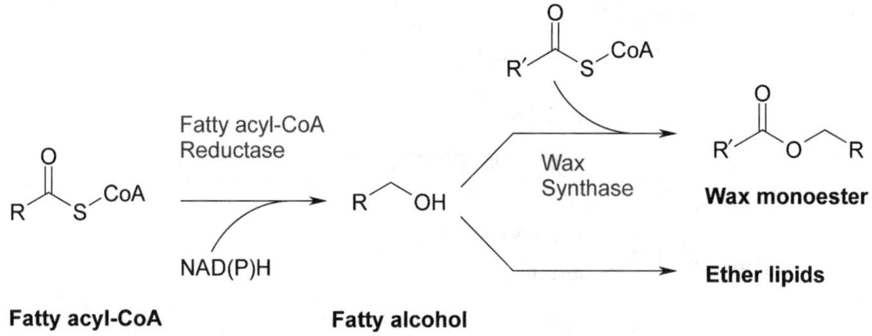

Figure 2.7 Scheme for the production of wax esters and enzymes in mammals and other organisms. Redrawn from Cheng and Russell.[19]

These enzymes must exist in several organisms as cDNAs specifying fatty acyl-CoA reductases have been identified in the jojoba plant, the silkworm moth, wheat and a micro-organism.[19]

Fatty alcohols have two metabolic fates in mammals: incorporation into ether lipids or incorporation into waxes. Ether lipids account for $\sim 20\%$ of phospholipids in the human body and are synthesised in membranes by a pathway involving at least seven enzymes. The second step of this pathway is catalysed by the enzyme alkyl-dihydroxyacetone phosphate synthase, which exchanges a fatty acid in ester linkage to dihydroxyacetone phosphate with a long chain fatty alcohol to form an alkyl ether intermediate. Once produced, ether lipids are precursors for platelet activating factor, for cannabinoid receptor ligands and for essential membrane components in cells of the reproductive and nervous systems.[19]

Metz *et al.*[17] have summarised the biochemistry of fatty alcohol synthesis; the process has been examined in diverse organisms and it has been demonstrated that the alcohols are formed by a four-electron reduction of fatty acyl-CoA[20–24] using a FAR enzyme. Although the alcohol-generating FAR reactions proceed through an aldehyde intermediate (Figure 2.8), a free aldehyde is not released.[25] Thus, the alcohol-forming FARs are distinct from those enzymes that carry out two-electron reductions of fatty acyl-CoA and yield free fatty aldehyde as a product.[26–28] A further distinction is that the alcohol-forming FARs are thought to be integral membrane proteins, whereas those that carry out two-electron reductions are either soluble enzymes or have a peripheral membrane association.

The range of fatty alcohols produced by organisms is, therefore, dependent on the fatty acids produced by the organism and the position within the synthesis pathway that the FAR reactions take place. The relative lack of polyunsaturated fatty alcohols indicates that these reactions take place before plants convert the unsaturated long chain acids to polyunsaturated acids (Figure 2.9).

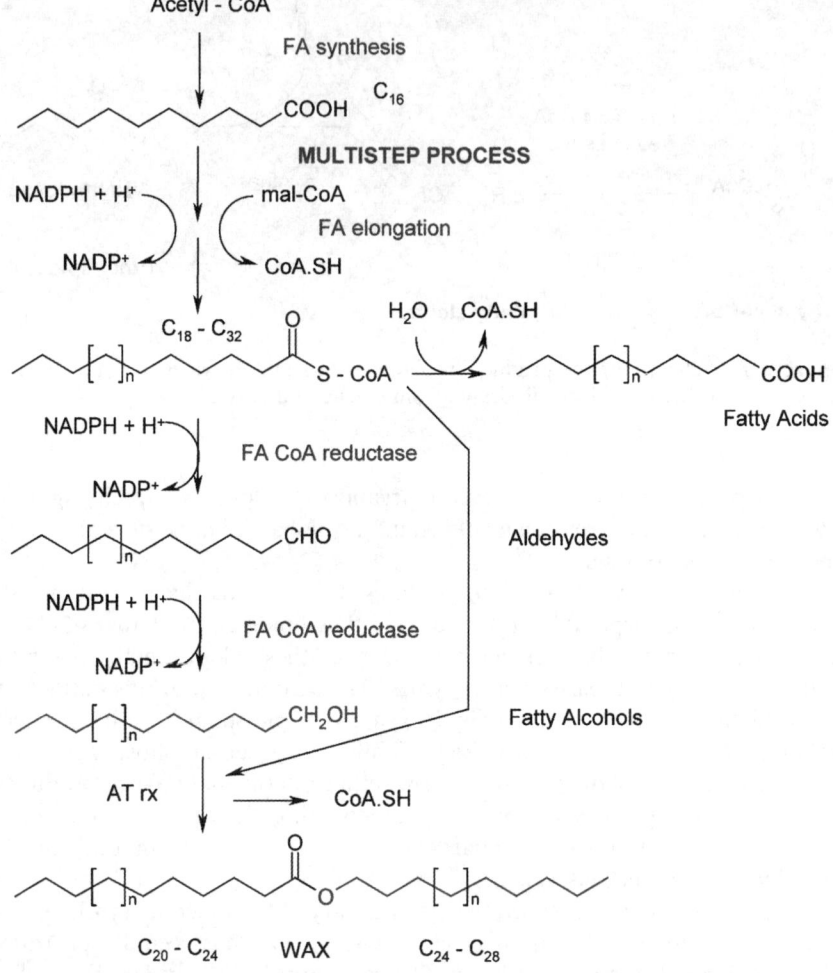

Figure 2.8   Scheme for fatty alcohol and wax production. Type I FAS produces a C_{16}
fatty acid that undergoes repetitive C_2 chain elongation. These may be
converted to "free" fatty acids by cleavage of the CoA.S group. Alternatively, the two-step FAR process converts the carboxylic acid group to
(i) an aldehyde and (ii) an alcohol. In many organisms, the acid and
alcohol are combined to produce a long chain wax.

2.5   Synthesis From Carbohydrates (Copepods)

There have been several studies of lipids in copepods, a small zooplankton
abundant in cool and temperate waters.[2,29–32] In general, copepods are heaviest
and rich in lipid shortly after the spring phytoplankton bloom. It has been
implied that these organisms make fatty acids and alcohols directly from a
carbohydrate source rather than by *de novo* synthesis from acetyl subunits. The
fatty acid and alcohol compositions of two *Calanus* species showed high levels

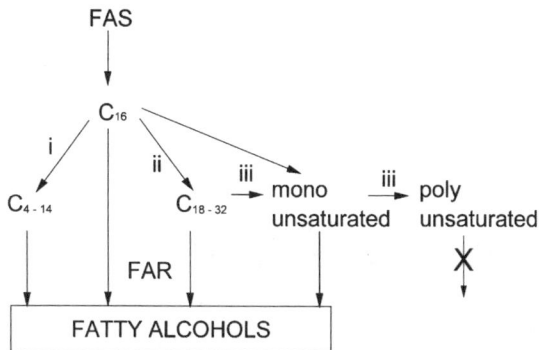

Figure 2.9 Schematic process for the formation of fatty alcohols from fatty acids. Reaction (i) is chain shortening by fatty acyl-CoA dehydrogenase; reaction (ii) is chain elongation by continued malonyl-CoA addition in plants; reaction (iii) is desaturation principally in the Δ^9, Δ^{12} and Δ^{15} positions, the latter two occurring in plants only.

of C_{16} acids and 20:5 acid, which are characteristic for diatoms.[33–35] A comparison of particulate matter in the sea with the data from *Calanus finmarchicus* in spring shows that the copepod fatty acids may originate directly from the particulate material, which consists of diatoms and a substantial amount of detritus.[30]

Summary

- Since fatty acids are the precursors of fatty alcohols, the profile in organisms tends to reflect the acid profile, although there are few polyunsaturated fatty alcohols.
- Type I FAS takes place in animals, fungi and some bacteria and uses a single large polypeptide for the elongation in two carbon subunits to C_{16} (palmitate).
- Type II FAS produces the same end product but the polypeptides are independent of each other and there may be other short chain products.
- Fatty alcohols tend to be made from fatty acids through the FAR system. Although this is a two-step system, free aldehydes are rarely produced.
- Branched chain and odd carbon number fatty acids and hence alcohols are synthesised almost exclusively by bacteria.
- Although many organisms make polyunsaturated fatty acids with two or more double bonds, these bonds are relatively rare in fatty alcohols.

CHAPTER 3
Occurrence in Biota

How much, where and why?

Due to the diversity of synthetic pathways outlined in Chapter 2, different organisms will contain or excrete different ranges of fatty alcohols. It is this diversity that has a key bearing on the ability to use fatty alcohols as bio-markers for different organic matter sources (see Chapter 8).

Most fatty alcohols occur in biota as waxes, esters with fatty acids (Figure 2.7). These compounds can have relatively high carbon contents. Chain lengths up to C_{64} have been observed.[36] These compounds serve several purposes to the organism that produces them. In the marine environment, these can be summarised in the following list:[32]

1. Energy reserve
2. Metabolic water reserve
3. Buoyancy generator
4. Biosonar lens in marine mammals
5. Thermal insulator.

In the terrestrial environment, another range of functions can be ascribed to the waxes from plants[37] and insects:[38,39]

1. Prevention of desiccation
2. Protection from bacterial attack
3. UV screening.

In birds, waxes are secreted by the uropygial or the preening gland and are used to maintain the condition of the feathers.[40] This may be for waterproofing as well as deterring pests.

Fatty Alcohols: Anthropogenic and Natural Occurrence in the Environment
By Stephen M Mudge, Scott E Belanger, and Allen M Nielsen
© Copyright 2008 ERASM (the joint surfactant environmental research platform of AISE and CESIO) and SDA
Published by the Royal Society of Chemistry, www.rsc.org

3.1 Bacteria

Bacteria synthesise a range of relatively short chain fatty acids and alcohols. Using Type II FAS, they synthesise compounds with chain lengths normally up to C_{18}.[41] However, due to their ability to use proprionyl-CoA and amino acids as the starter for FAS, these compounds can have odd as well as even chain lengths.[4] The orientation of this starter also leads to the formation of *iso* and *anteiso* branched chain compounds as well. Typical compounds are in the range of C_9 to C_{20}.[42]

Johns *et al.*[43] state that "branched chain fatty acids in particular have been considered to reflect a bacterial origin". Both *iso* and *anteiso* branched chain acids are common to many bacteria[44–46] and account for up to 60% of the total fatty acids in many *Bacillus* species.[13] Due to the synthetic pathway for fatty alcohols, fatty acids should act as a good indicator of the likely fatty alcohols found in bacteria. Unlike plants, in bacteria the fatty acids are part of the cell membrane but the role of fatty alcohols is not known.

The occurrence of wax esters in bacteria of the genus *Acinetobacter* is principally as an energy storage reserve.[47] In bacteria, waxes are accumulated as cytoplasmic inclusions surrounded by a thin boundary layer similar to eukaryotes. Recently, a wax ester synthase/acyl-CoA:diacylglycerol acyl transferase (WS/DGAT) was identified from *Acinetobacter calcoaceticus* which catalysed the key steps in the biosynthesis of both storage lipids.[48] A large number of WS/DGAT-related proteins were identified in the genome sequences of triacylglycerol (TAG) and wax ester accumulating bacteria like *Mycobacterium tuberculosis, M. leprae, M. bovis, M. smegmatis* and *Streptomyces coelicolor*; it may be assumed that this type of enzyme is responsible for wax ester and TAG biosynthesis in all oleogenous bacteria.[49]

It has been suggested that the C_{14}–C_{18} distribution of fatty acids in sediments reflects an input from bacteria.[50] Further evidence for bacterial activity exists in the presence of *iso* and *anteiso* C_{15} and C_{17} acids.[51–54] Work on these fatty acids present in marine sediment[50] provides a good indicator of the likely fatty alcohol series that might be seen as well; few data are available directly on the fatty alcohols in bacteria. In their experiments, Parkes and Taylor[50] identified several short chain acids, several of which were also unsaturated. The profile of the straight chain and *iso* and *anteiso* branched compounds can be seen in Figure 3.1. The dominant compound is 16:0 in all cases, but substantial amounts of odd chain and branched compounds are also present. Parkes and Taylor[50] suggest that the *anteiso* C_{15} may be indicative of sulfate reducing bacteria (SRB).

3.2 Chlorophyll Side Chain (Phytol)

One of the major fatty alcohols in the environment is the phytol molecule derived from the side chain of chlorophyll[55,56] (Figure 1.3). Chlorophyll, the major photosynthetic pigment of green plants, is comprised of a tetrapyrrole

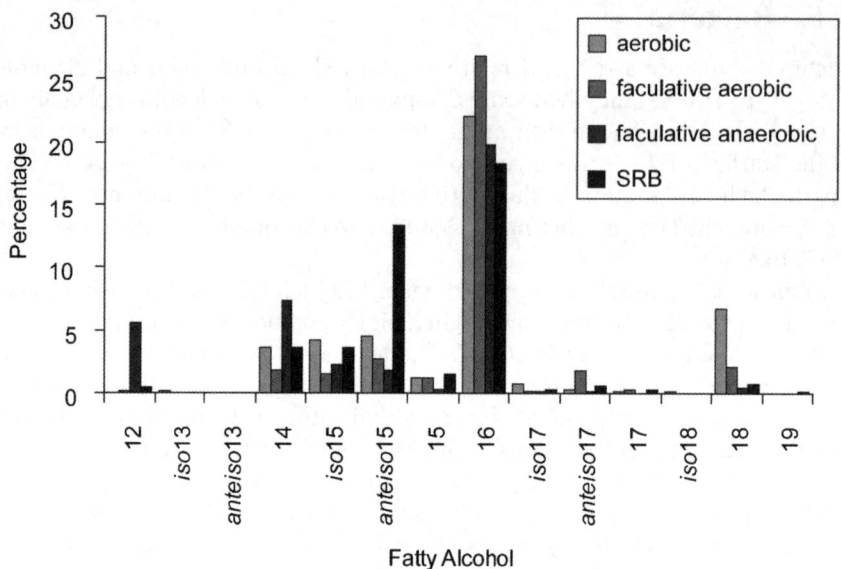

Figure 3.1   Percentage straight chain and *iso/anteiso* branched fatty acids in different types of marine bacteria. Data from Parkes and Taylor.[50]

ring structure coordinating a magnesium atom. This part of the molecule harvests the photons of incident radiation and passes it along an electron transport system. The phytyl side chain is mainly present to impart a degree of hydrophobicity to reduce the water solubility and immobilise the chlorophyll within the cells. The synthesis of the phytyl side chain is from an isoprenoid system using mevalonic acid and does not rely on a fatty acid precursor.[1]

Analysis of environmental samples by saponification (see Chapter 6) will release the phytol from chlorophyll into the solvent. Therefore, the phytol may be a good indicator of the chlorophyll in the sample. This may originate in both the terrestrial environment from green plants and in the marine environment from phytoplankton. However, recent ^{13}C stable isotope analysis of marine, estuarine and terrestrial samples shows that the phytol in the marine and estuarine samples was entirely derived from the marine environment and not terrestrial plants (δ^{13}C value of −23‰ compared to −33‰ for land samples; S.M. Mudge, unpublished data). Therefore, if there is little transfer from the land to the sea, the phytol may be used as an indicator of primary productivity in the marine environment. Further information on the use of stable isotopes can be found in Chapter 8.

Free phytol can also be generated in the water column by hydrolysis of chlorophyll,[57] during the digestive processes of copepods feeding on phyto-plankton[58] and by senescence of diatoms.[59] Experiments with senescent algae have shown that, under high light conditions, chlorophyll can degrade to form a diol after alkaline hydrolysis (Figure 3.2).[60] This diol has been detected in

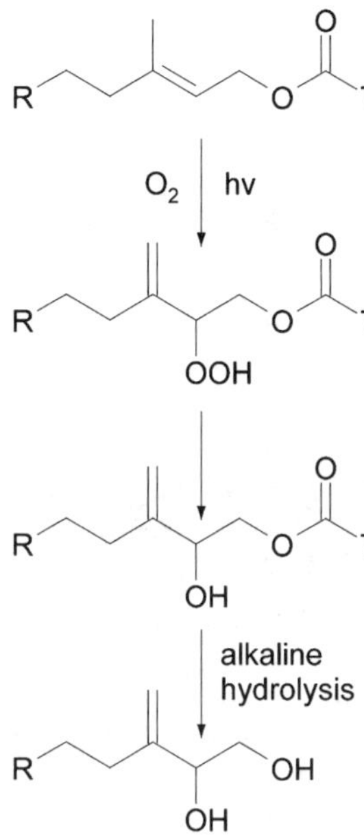

Figure 3.2 Phytyl diol production from chlorophyll under high light conditions. R, rest of the phytyl side chain of chlorophyll (see Figure 1.3); TP, tetrapyrrole structure of chlorophyll. Modified from Cuny and Rontani.[60]

many sediment samples, including those in deep cores (S.M. Mudge, unpublished data) suggesting the method may be appropriate to the reconstruction of historical solar irradiation values.

3.3 Marine Plants

Marine plants do not have substantial wax concentrations; there are few reports of long chain alcohols in microalgae and it appears that microalgae are not a major source of these lipids in most sediments.[61]

Some marine plants do have some fatty alcohols and these have been reviewed by Volkman *et al.*[61] For example, an 18:1 fatty alcohol has been found in the diatom *Skeletonema costatum.*[62] A series of C_{22}–C_{28} saturated *n*-alcohols, with even carbon numbers predominating, and a maximum at C_{26} and C_{28}, has been identified in the heterocyst glycolipids of the cyanobacterium *Anabaena cylindrical.*[63] The green alga *Chlorella kessleri* contains C_{10}–C_{20} saturated and mono-unsaturated fatty alcohols, with 16:0 most abundant, all esterified to long chain fatty acids.[64,65] C_{30}–C_{32} alcohols having one or two double bonds are significant constituents of the lipids of marine eustigmatophytes of the genus *Nannochloropsis.*[66] The freshwater eustigmatophytes *Vischeria punctata* contains saturated and mono-unsaturated *n*-alkanols from C_{16} to C_{28} showing a strong predominance of 22:0 and 26:1, respectively. It is reported that in the

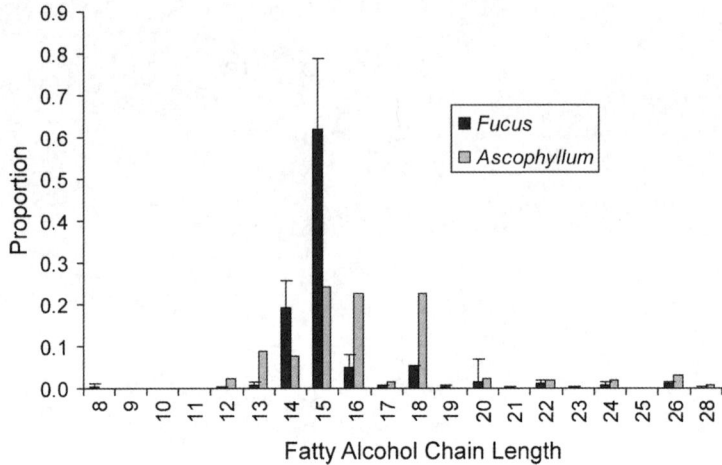

Figure 3.3   Fatty alcohol profile for *Ascophyllum nodosum* and *Fucus sp.* collected from an upper intertidal marine area. Unpublished data from Le Crom and Mudge.

alcohol data there is not a steady decline in abundances with increasing chain length, but rather a strong predominance of just a few homologues.[61]

There are no published reports of series of *n*-alkanols being found in marine macroalgae. The only report of interest is of a polyunsaturated compound found in the red alga *Gracilaria foliifera* which would not be identified in traditional environmental analyses.[67] However, analyses were conducted on the brown algae *Ascophyllum nodosum* and *Fucus spiralis* for this book and the profile of the measured alcohols can be seen in Figure 3.3. The profile is quite surprising in that the C_{15} alcohol dominates and there are high concentrations of odd chain compounds in the C_{12}–C_{18} range. The fronds had been washed under UV treated filtered seawater prior to analysis and, with the *Fucus* samples, had come from six locations in a transect from low water springs to the high water mark. This profile would be indicative of bacteria and they may originate from a surface coating of the algal fronds rather than a waxy surface layer. There are a few long chain moieties up to C_{28} present but these are very much minor constituents compared to the C_{13} to C_{18} compounds.

3.4   Terrestrial Plant Waxes

The main function of plant waxes is to reduce water loss through evaporation. Therefore, the chain length of the waxes for this source tends to be longer than that for marine animals. Typical profiles of fatty alcohols for selected plants can be seen in Figures 3.4 and 3.5. As can be seen in these figures, short chain alcohols ($<C_{20}$) are not present and the most prominent chain lengths are C_{26} to C_{30}. In the example from the African grasslands,[68] compounds up to C_{34} are

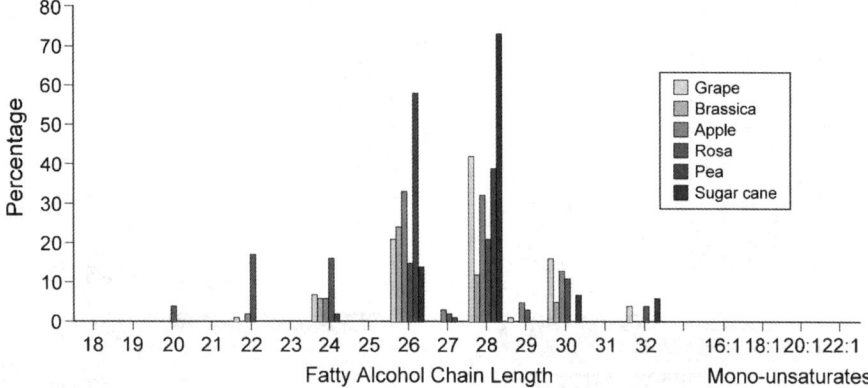

Figure 3.4   Fatty alcohols from terrestrial plants. The major fatty alcohols occur either at C_{26} or C_{28} with no mono-unsaturated compounds. After Tulloch.[36]

relatively common. In that study, odd chain alcohols were present but at very low concentrations and were not reported. There is also an even over odd dominance indicative of the C_2 (alkyl) chain elongation process. Few unsaturated fatty alcohols have been reported in plants, unlike marine animals. The longer chain length compounds in the African plant samples compared to the temperate plants in Figure 3.4 suggest that chain length may be directly related to adaptive behaviour of the plants to control evaporative losses.

3.5   Mosses and Other Peat-Forming Plants

Some terrestrial plants live in wet areas and when they die their remains degrade *in situ* and may form peat. Typical plants of these environments are the mosses, and a study by Nott[69] investigated the fatty alcohol and other biomarker composition of mosses as part of a larger study of Bolton Fell Moss[70] (see also Chapter 7). The profile of fatty alcohols in several moss species and other peat-forming plants can be seen in Figures 3.6–3.8. The *Sphagnum* species (Figure 3.6) have very similar profiles although the other plants are less similar. The major alcohols are in the C_{24}–C_{32} range with very few odd chain compounds.

3.6   Marine Animals

In contrast to terrestrial plants, marine animals principally use fatty alcohols as a storage product.[32] Examples from several different organisms are shown in Figure 3.9. No saturated fatty alcohols with chain lengths greater than C_{19} were observed, although a substantial amount of mono-unsaturated compounds were present. As with terrestrial plants, few odd chain length compounds were present in the samples.

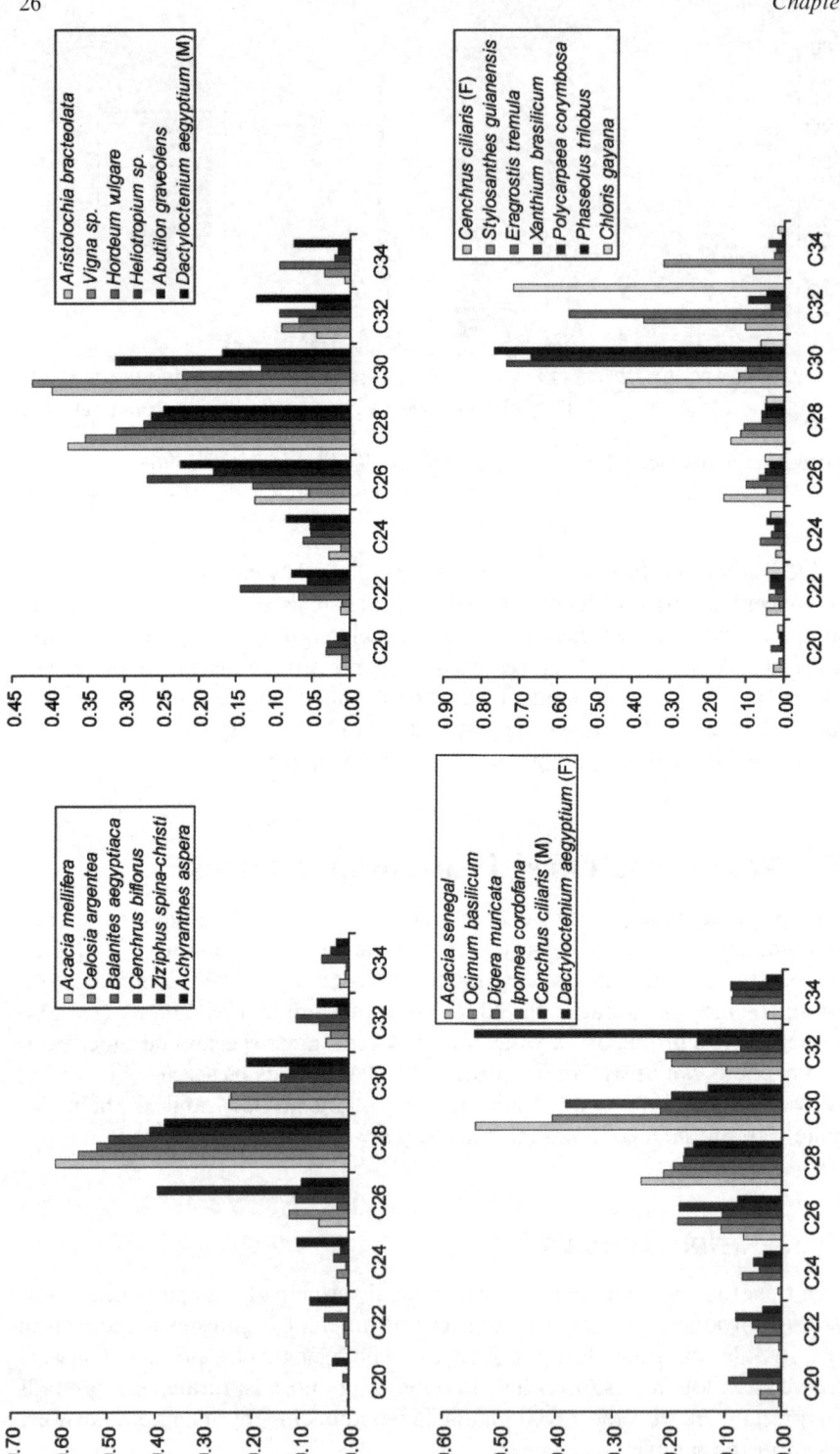

Figure 3.5 Fatty alcohol chain length profiles (as proportions) for a range of terrestrial plants from African grassland.[68] Note the differing scales on the *y*-axes. M, mature plant; F, flowering plant.

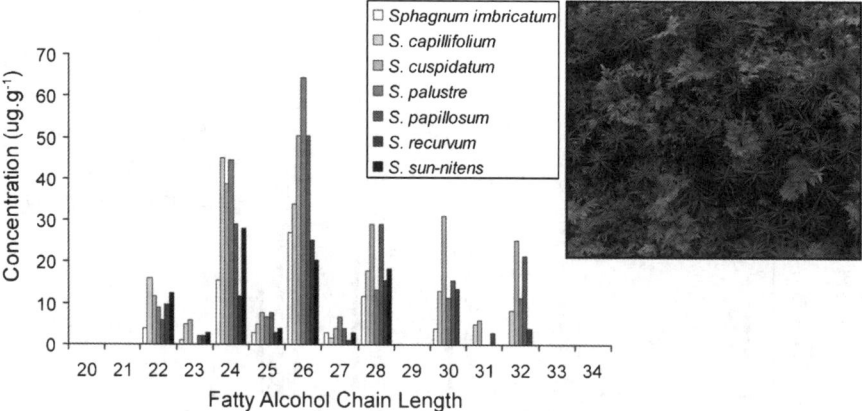

Figure 3.6 Fatty alcohol chain length profile of mosses from the genus *Sphagnum*. Samples were collected from Bolton Fell Moss. Data after Nott.[69] Examples of typical mosses are shown in the inset.

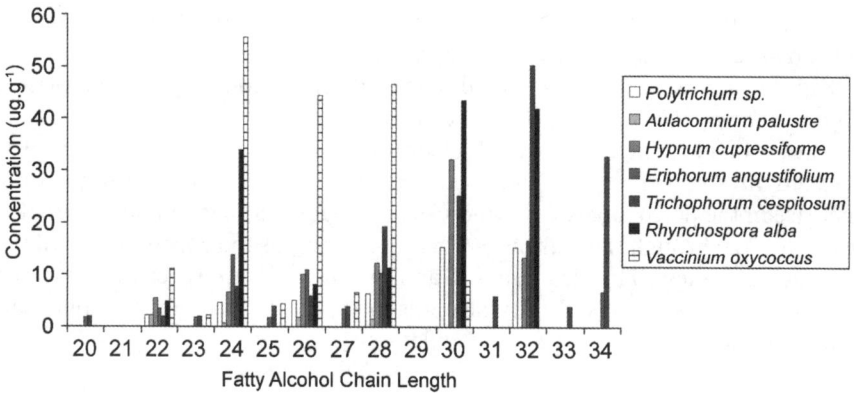

Figure 3.7 Fatty alcohol chain length profile of peat-forming plant species. Samples were collected from Bolton Fell Moss. Data after Nott.[69]

Work by Hamm and Rousseau[71] on the foam found after a *Phaeocystis* bloom showed that it consisted of particulate and dissolved matter that did not contain the typical algal pattern of *Phaeocystis*-derived fatty acids. They found a high abundance of fatty alcohols, which are indicators of wax esters and, thus, zooplankton in the foam;[72] they thought this was surprising since apart from a few copepod eggs no zooplankton was found in the foam. Hamm and Rousseau suggested that the fatty alcohols may have originated from dissolved zooplanktonic wax esters, a phenomenon which has been observed[72] within a "milky water" event in the North Sea. Wax esters are thought to occur in high concentrations in over-wintering copepods (as a food reserve[32]), but less so in vernal zooplankton.[31] However, Hamm and Rousseau[71] demonstrated that

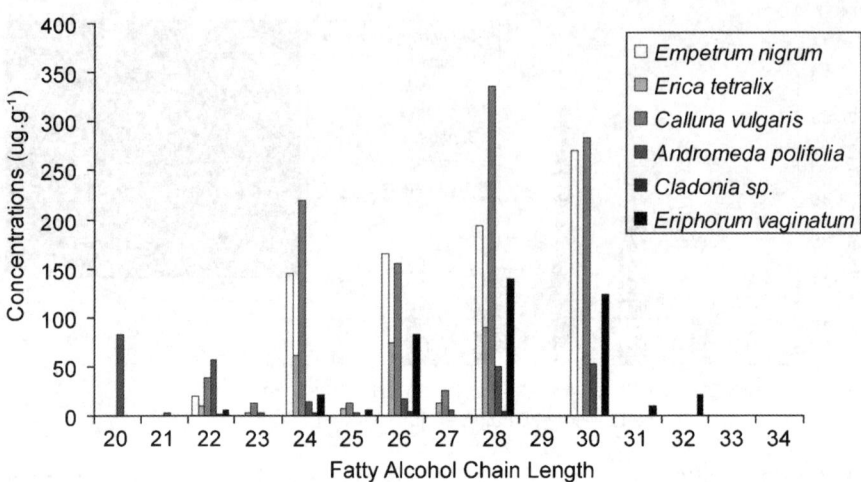

Figure 3.8   Fatty alcohol chain length profile of peat-forming plant species. Samples
were collected from Bolton Fell Moss. Data after Nott.[69]

high percentages (20–92%) of wax esters are found in the lipids of *Calanus
finmarchicus* in all stages of its development.

Kattner and Krause[30] also found a seasonal variation between samples of
Pseudocalanus elongatus; those collected in spring had a relatively high percentage of short chain saturated alcohols ($C_{14} + C_{16} = 87\%$) but this was
reduced in summer (60%) and winter (30%). There was a corresponding
increase in the percentage of C_{20} and C_{22} mono-unsaturated compounds (6% –
26% – 69%) through the seasons (spring to autumn) as the copepods stored the
carbon as waxes. This led Hamm and Rousseau[71] to speculate that the
occurrence of the dissolved fatty alcohols in the post-*Phaeocystis* bloom indicated the mortality of a copepod population.

3.7   Insects

The cuticular surfaces of insects are also covered by a lipid layer whose primary
function is to restrict water movement across the cuticle, preventing desiccation
of the insect.[38] The major components in the cuticular extracts of insects
include hydrocarbons, wax esters, aldehydes, ketones, alcohols and acids. The
quantities and composition of free cuticular lipid can vary widely among insect
groups and sometimes within the developmental stages of the same species. In a
study of two lepidopteran species, Buckner *et al.*[38] found similar fatty alcohol
compounds present in each, although in one species C_{26} had the maximum
occurrence while it was C_{30} in the other (Figure 3.10). In both cases, the
composition is similar to that of terrestrial plants on which the species feed.

In general, homopteran insects produce large amounts of wax.[39] This wax is
in the form of filaments or particles in many whitefly species. These particles

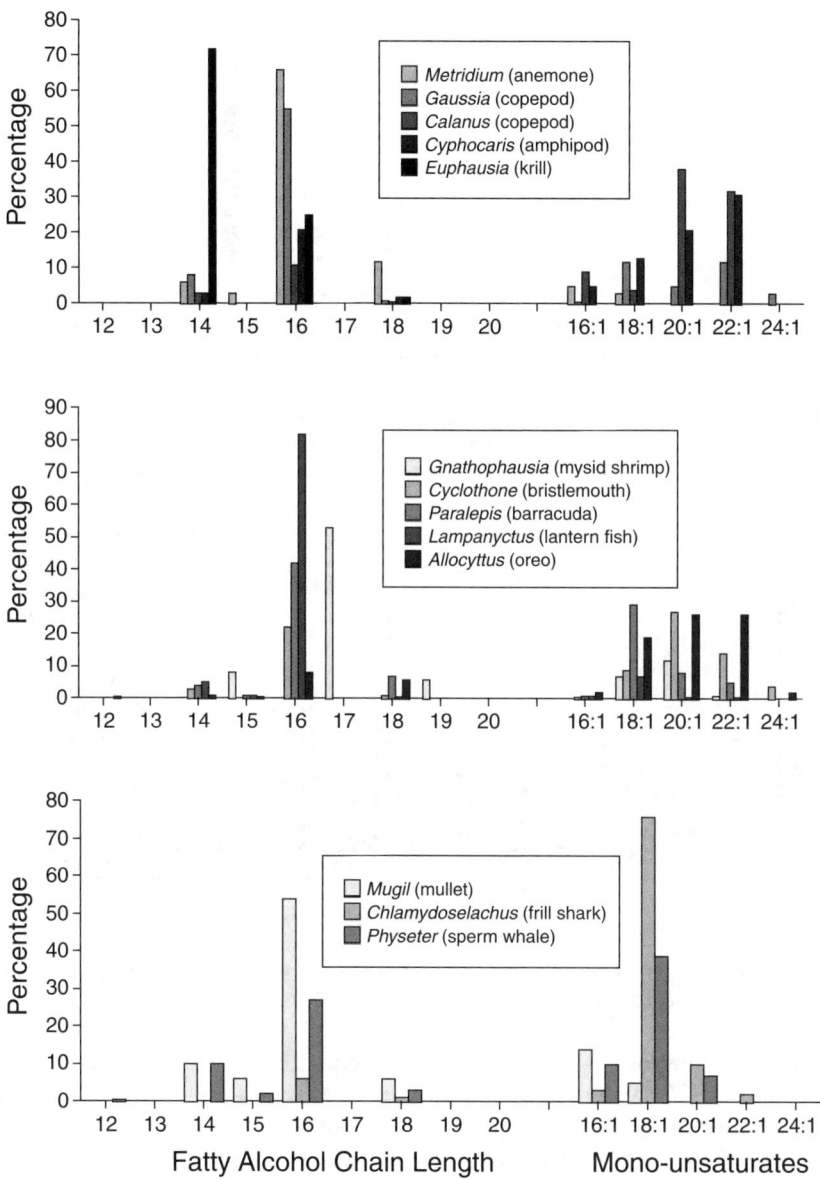

Figure 3.9 Fatty alcohol chain length profiles from several types of marine animals. The fatty alcohols are principally short chain with a maximum carbon chain length of 16.

have been identified as not being composed of wax esters but of a mixture of a long chain aldehyde and a long chain alcohol; for example, the greenhouse whitefly (*Trialeurodes vaporariorum*) has dotriacontanal (C_{32} aldehyde) and dotriacontanol (C_{32} alcohol). In the sweet potato whitefly (*Bemisia tabaci*) the

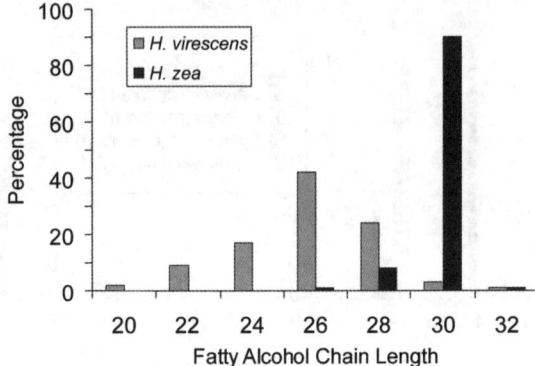

Figure 3.10 Fatty alcohol chain length composition in the wax esters from two lepidopteran species: the tobacco budworm *Heliothis virescens* and the corn earworm *Helicoverpa zea*. Data after Buckner *et al.*[38]

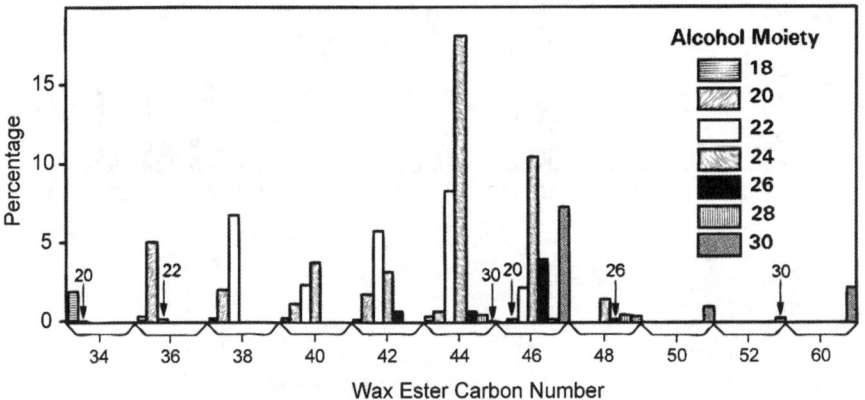

Figure 3.11 Fatty alcohol chain length composition of wax esters in the giant whitefly *Aleurodicus dugesii*. Redrawn from Nelson *et al.*[39]

waxes are composed of the C_{34} equivalents tetratriacontanal and tetra-triacontanol. The external wax of whitefly nymphs may play a role in the parasitisation or predation of nymphs which are often preferred prey over adult whiteflies. In a study of the external lipids of the giant whitefly (*Aleurodicus dugesii*) Nelson *et al.*[39] found a range of alcohols present within the adult wax esters (Figure 3.11). The peak chain lengths are similar to those found in terrestrial plants (C_{20}–C_{30}).

3.8 Birds

Birds use waxes to maintain waterproofing of their feathers and resist microbial/fungal infestation. The preen gland (uropygial gland) is located under the

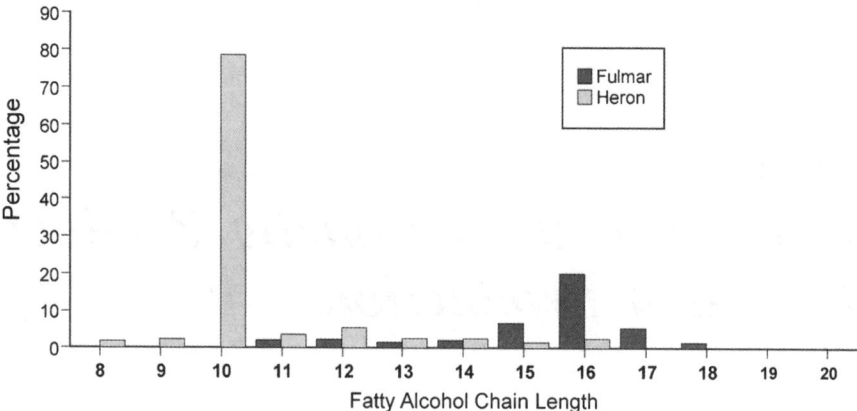

Figure 3.12 Fatty alcohol chain lengths from the preen gland of selected birds. After Jacob.[40]

tail feathers and secretes a simple wax derived from fatty acids and fatty alcohols. The typical profiles for the alcohols in these waxes can be seen in Figure 3.12. Usually, many of the fatty alcohols from this source tend to have methyl branches in the 2, 4, 6 or 8 position.[40]

Summary

- Bacteria tend to produce short chain even carbon number fatty alcohols together with a suite of odd carbon chain length and *iso* and *anteiso* branched compounds.
- Terrestrial plants produce long chain even carbon number fatty alcohols which are esterified to fatty acids to produce waxes to reduce desiccation.
- Marine animals do not have the same problem of water loss and synthesise short chain even carbon number compounds. Birds tend to have the same profile.
- Insects have fatty alcohol profiles similar to those of the plants on which they feed and so contain long chain even carbon number compounds.

CHAPTER 4

Consumer and Cosmetic Product Uses and Production

As well as the natural production of fatty alcohols, they are manufactured for incorporation into consumer products: how is this achieved and what do these formulations contain?

Fatty alcohols in the C_6 to C_{22} range are used in the manufacture of some major classes of ionic and anionic surfactants. Approximately 50% of the manufactured fatty alcohol volume is consumed in these uses. The remainder of the volume of the manufactured fatty alcohols finds a wide range of applications utilising their lubricating, emollient, solubilising or emulsifying properties. They are widely applied in industrial applications and can found in many products for consumer and professional use based on their lubricating, solubilising or emulsifying properties. They are used in certain paints, lubricants, emulsifiers, flotation agents, and rolling and formwork oils. They are also used as an additive in certain plastics, paper products and plaster, and are used in processing of textiles, leather and plastics. Fatty alcohols are also present in some pharmaceutical products and agrochemical formulations.

The majority of fatty alcohols and their derivatives are used in consumer products such as household cleaning products (*e.g.* liquid and powder detergent surfactants, general and hard surface cleaners, fabric conditioners). Their use in personal care products includes shampoos, hair conditioners, styling gel and mousse, cleansers, body washes, skin lotions and creams, antiperspirants, face and eye cosmetics, make-up remover and hair dyes. They may also be used in fragrances and fragrance ingredients. In most cases, fatty alcohols make up 1–5% of the final product formulation. The exceptions to this are solid antiperspirants which may contain up to 25% fatty alcohols (Table 4.1).

Fatty Alcohols: Anthropogenic and Natural Occurrence in the Environment
By Stephen M Mudge, Scott E Belanger, and Allen M Nielsen
© Copyright 2008 ERASM (the joint surfactant environmental research platform of AISE and CESIO) and SDA
Published by the Royal Society of Chemistry, www.rsc.org

Table 4.1   Percentage fatty alcohol content in a range of product formulations used in the home.

| Product category | Product formulations by region[a] (% fatty alcohols) | | | | | |
| | USA | | EU | | AP | |
	Mean	Range	Mean	Range	Mean	Range
Household products						
Laundry powder	3	1–5	3	1–5	3	1–5
Fabric softener	3	1–5	3	1–5	0.75	0.5–1
Surface cleaners	3	1–5	–	–	3	1–5
Personal care and cosmetic products						
Hair conditioners	3	1–5	3	1–5	1.88	1.5–5
Hair mousse					3	1–5
Hair dyes					7.5	5–10
Skin lotions	3	1–5	3	1–5	1.88	0.5–5
Antiperspirants (solid)	17.5	10–25	17.5	10–25	17.5	10–25
Face/eye liquid	3	1–5	3	1–5	3	1–5
Face/eye powder	3	1–5			0.75	
Mascara					3	1–5

[a]USA, United States of America; EU, European Union; AP, Asia Pacific Region.

Fatty alcohols are used in several industrial formulations utilising their lubricating or emulsifying properties (*e.g.* flotation agents, lubricants, emulsifiers, formwork oils, rolling oils, oilfield chemicals and defoamers).

4.1   Detergent Alcohols Manufacture

Alcohols used in detergent applications are typically 12 or more carbons in length with a maximum in the C_{12}–C_{18} range. These alcohols are commercially produced in a number of ways, but the resulting products are usually classified according to the source of raw materials used to produce them. There are two general categories: those derived from biological fats and oils (*oleochemical*) and those derived from crude oil, natural gas, natural gas liquids or coal (*petrochemical*).[73]

4.1.1   Oleochemical-based Alcohols

In natural fats and oils, the hydrocarbon chains have already been formed in the raw material. Biological processes in living organisms synthesise long carbon chains (see Chapter 2), often in the form of triacylglycerol (triglycerides). In the manufacture of feedstock for subsequent incorporation into final formulations, the fatty acids from plant and animal oils and fats are separated from the triglycerides and chemically converted into key alcohol intermediates. Coconut oil and palm kernel oil are preferred for the production of C_{12}–C_{14} chain lengths; animal fats (tallow) and palm oil are preferred for the production of C_{16}–C_{18} chain lengths. A breakdown of selected potential sources of these fatty acids is shown in Table 4.2.[73,74]

Table 4.2 Fatty acid composition (wt%) of natural triglycerides.

Triglyceride fat or oil	Caprylic 8:0	Capric 10:0	Lauric 12:0	Myristic 14:0	Myristoleic 14:1ω5	Pentadecanoic 15:0	Palmitic 16:0	Palmitoleic 16:1ω7	Margaric 17:0	Stearic 18:0	Oleic 18:1ω9	Linoleic 18:2ω6	Linolenic 18:3ω3
Tallow				3.2	1.0	0.4	26.4	2.6	0.9	26.9	36.7	(1)	
Palm				0.9			46.6			4.1	39.3	9.1	
Coconut	8.0	6.7	51.3	16.2			7.6			2.7	5.9	1.6	
Palm kernel	4.0	5.0	50.0	15.0			7.0	0.5		2.0	15.0		1.0

4.1.1.1 Fatty Acids

In general, the ester linkage in triglyceride molecules can be severed in two ways. In one process, steam is used to hydrolyse the triglycerides to yield fatty acids and glycerol[73] (Figure 4.1).

4.1.1.2 Fatty Acid Methyl Esters (FAMEs)

In another process, methanol is used to transesterify triglycerides to yield fatty acid methyl esters and glycerol[73] (Figure 4.2). In both cases, the glycerol produced as a by-product can be collected and used in other formulations.

4.1.2 Oleochemical Fatty Alcohols

Oleochemical fatty alcohols of C_{12} to C_{18} chain lengths are produced by the hydrogenation of both FAMEs and fatty acids according to the methods shown in Figures 4.3 and 4.4.[74]

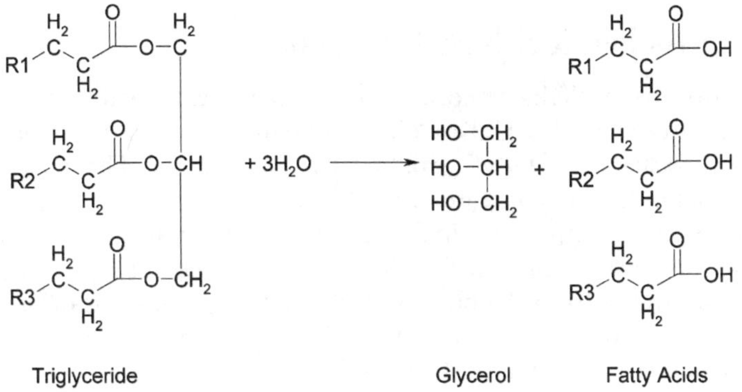

Figure 4.1 Fatty acid production (fat splitting) from triglycerides.

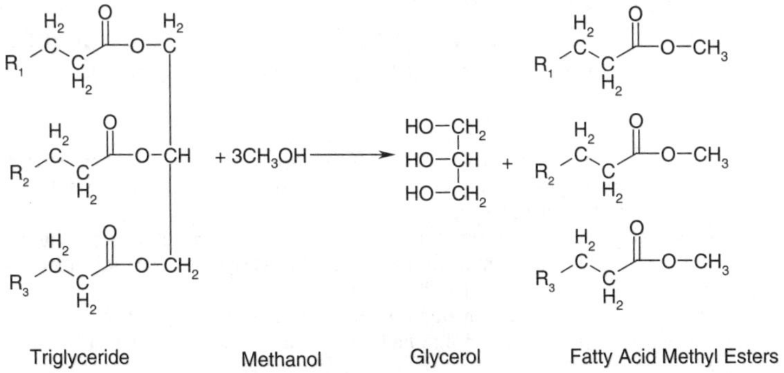

Figure 4.2 Fatty acid methyl ester (FAME) production.

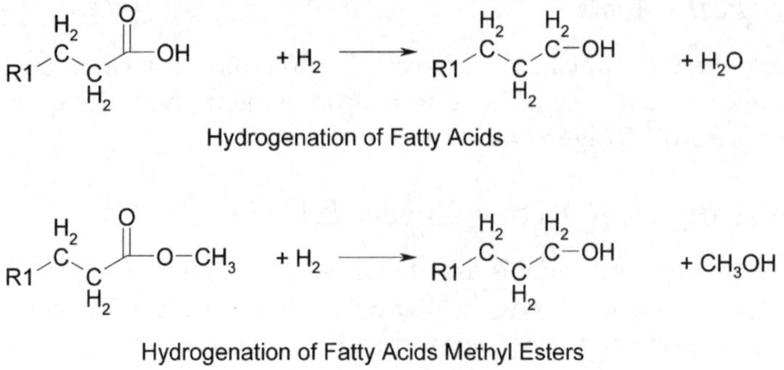

Hydrogenation of Fatty Acids

Hydrogenation of Fatty Acids Methyl Esters

Figure 4.3 Oleochemical alcohol production from fatty acids and FAMEs.

These alcohols have even carbon chain lengths, $>99\%$ linear, and are primary alcohols reflecting their biological source. Fatty alcohols are important oleochemical-based surfactant intermediates; many surfactant products are made from them, including alcohol sulfates, alcohol ethoxylates and alcohol ether sulfates.

4.2 Petrochemical-based Alcohols

Linear hydrocarbon chains or normal paraffin can be extracted from petroleum fractions. Kerosene and gas oil are different boiling fractions of petroleum that contain hydrocarbons of C_{10}–C_{16} and higher chain lengths. These may be used as precursors in the manufacture of alcohols.

Kerosene is an important hydrocarbon source. Using molecular sieve separation processes such as MOLEX[i] or ISOSIV[ii] the linear or normal paraffins are separated from the branched and cyclic hydrocarbons. The normal paraffin is distilled into various cuts and the branched/cyclic hydrocarbon stream or raffinate is sold as an upgraded fuel[73] (Figure 4.5).

4.2.1 Internal Olefins

Pure internal olefins can be produced from normal linear paraffin. In the combined PACOL/OLEX process, dilute PACOL[iii] olefins are concentrated by the OLEX[iv] process to about 96% internal olefins[73] (Figure 4.6).

---

[i] The MOLEX process from Universal Oil Products (UOP) uses a simulated moving bed with binderless zeolite as an absorbent and a light paraffin as a desorbent. Its prime use is the separation of n-paraffins from naphtha and kerosene.[75]

[ii] The ISOSIV process (ISOmer separation by molecular SIeVes) was developed by Union Carbide and is widely licensed by UOP. It is similar to the MOLEX process separating linear hydrocarbons from naphtha and kerosene in the vapour phase.[76]

[iii] The PACOL process, also licensed through UOP, converts paraffins to mono-olefins. This dehydrogenation is carried out over a noble metal with added H_2 at temperatures greater than 400 °C.[77]

[iv] OLEX is a continuous liquid chromatography separation using a zeolite as an adsorbent. The process was also developed by UOP.[76]

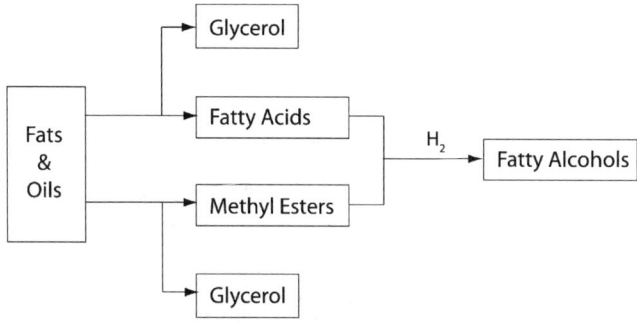

Figure 4.4 Summary of oleochemical alcohol production and glycerol as the main by-product. Glycerol maybe used in the food and beverage industry and in pharmaceutical formulations to improve smoothness.

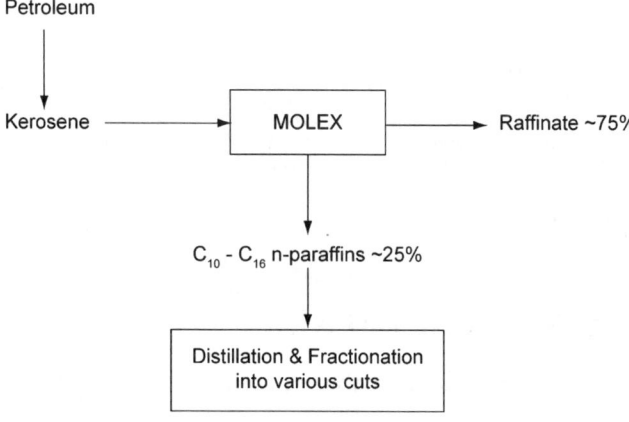

Figure 4.5 Normal paraffin production.

4.2.2 Conventional OXO Alcohols Based on Internal Olefins

Internal olefins can be converted to conventional OXO alcohols. In contrast to oleochemical-based alcohols, OXO alcohols have both odd and even carbon chain lengths and they have up to 50% branching at the second carbon position.[74] The OXO reaction, as applied to the synthesis of detergent-range alcohols, involves the reaction of olefins with synthesis gas (CO/H_2) in the presence of an OXO catalyst to yield higher alcohols. The sequence of steps includes the following: hydroformylation, catalyst removal and recycling, aldehyde distillation, aldehyde hydrogenation and purification of the product alcohols, as shown in Figure 4.7.[74,78]

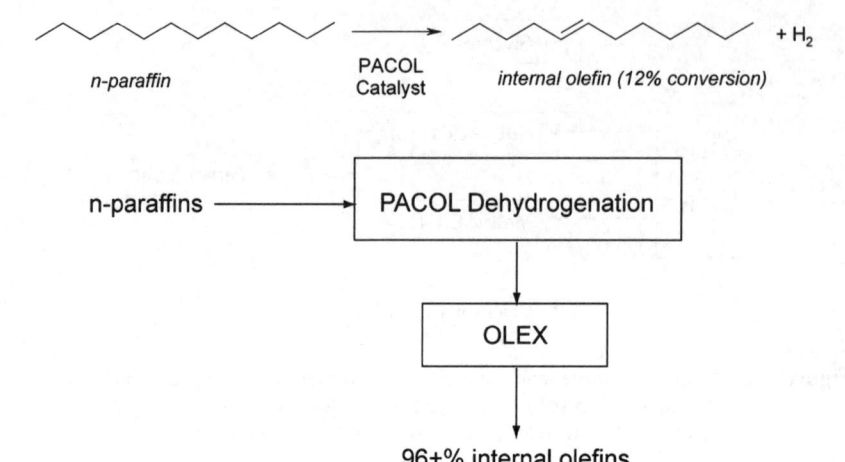

Figure 4.6   PACOL dehydrogenation of *n*-paraffins and the OLEX concentration route.

Figure 4.7   The OXO process.

4.3   Alcohols Based on Ethylene

4.3.1   Ziegler Ethylene Growth Process

Ethylene (ethene) can be used as a building block to form long hydrocarbon chains. This process employs a growth reaction to make hydrocarbon chains from C_2 to C_{20} in length. Hydrocarbon chains are grown by adding ethylene units to an organometallic compound such as triethylaluminium. The ethylene units are inserted between the growing alkyl chains and the aluminium, producing trialkylaluminium or growth product, as shown in Figure 4.8.[73]

4.3.2   Ziegler Alcohols

Further processing of the growth product yields linear primary alcohols. In the Ziegler alcohol process, linear even carbon chain fatty alcohols are produced from the growth product by controlled oxidation followed by hydrolysis. For a given chain length, these alcohols are essentially identical to natural alcohols, having linear even carbon chain length primary structures. A stoichiometric amount of aluminium is used in this process that eventually is converted into high-purity alumina after hydrolysis[73] (Figure 4.9).

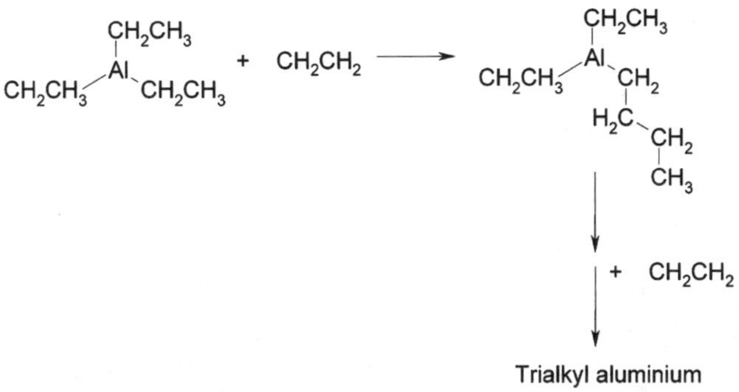

CH₂CH₃ / Al / CH₂CH₃ CH₂CH₃ + CH₂CH₂ ⟶ ... Trialkyl aluminium

Figure 4.8 Ziegler ethylene growth process.

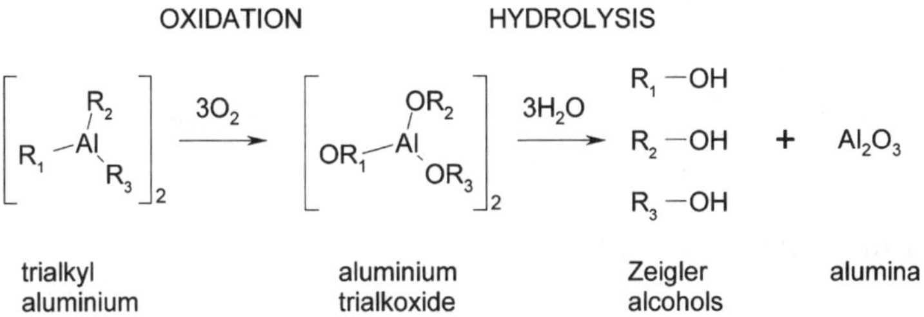

OXIDATION HYDROLYSIS

$$\left[\begin{array}{c} R_1-\underset{R_3}{\overset{R_2}{Al}} \end{array} \right]_2 \xrightarrow{3O_2} \left[\begin{array}{c} OR_1-\underset{OR_3}{\overset{OR_2}{Al}} \end{array} \right]_2 \xrightarrow{3H_2O} \begin{array}{c} R_1-OH \\ R_2-OH \\ R_3-OH \end{array} + Al_2O_3$$

trialkyl aluminium   aluminium trialkoxide   Zeigler alcohols   alumina

Figure 4.9 Ziegler alcohol process chemistry.

4.4 Modified OXO Alcohols

4.4.1 SHOP (Shell Higher Olefin Process) α-Olefins

The SHOP process employs an ethylene oligomerisation reaction to make α-olefins. In the first part of the process, linear even carbon chain α-olefins are produced. As with other ethylene growth reactions, the olefins are produced with a broad distribution of carbon chain lengths. Some of these chain lengths are more desirable as α-olefin products than others and are separated by distillation and sold in a range of molecular size fractions.[74]

4.4.2 SHOP Internal Olefins and Modified OXO Alcohols

In the second part of the SHOP process, α-olefins of the less desirable chain lengths are converted to linear internal olefins in a complicated process called isomerisation/disproportionation/metathesis. The internal olefins produced in this process have both odd and even chain lengths in the range C_{10}–C_{14}, as summarised in Figure 4.10.[74]

Internal olefins from the SHOP process are converted in the modified OXO process which produces alcohols having 20% branching[74,78] (Figure 4.11).

A modified OXO process exists that produces mono-methyl branched alcohols in the C_{16}–C_{17} range. The starting material in the upper left-hand side of

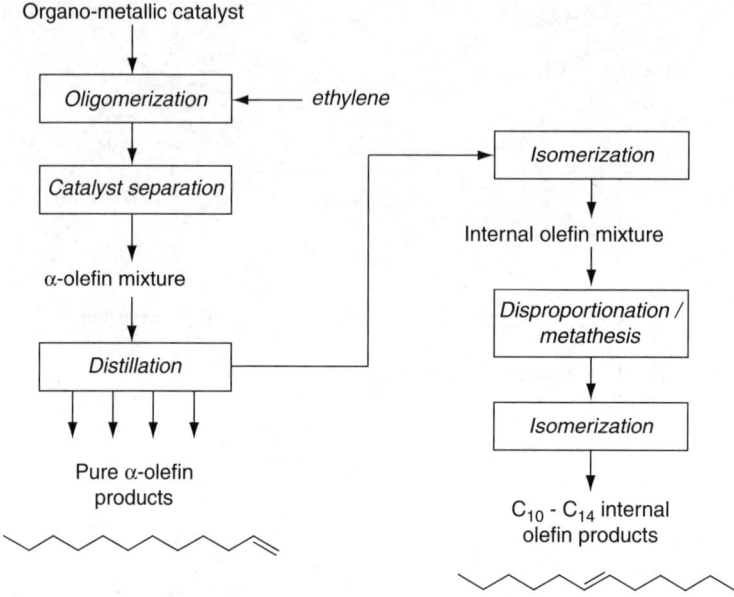

Figure 4.10   SHOP olefin process.

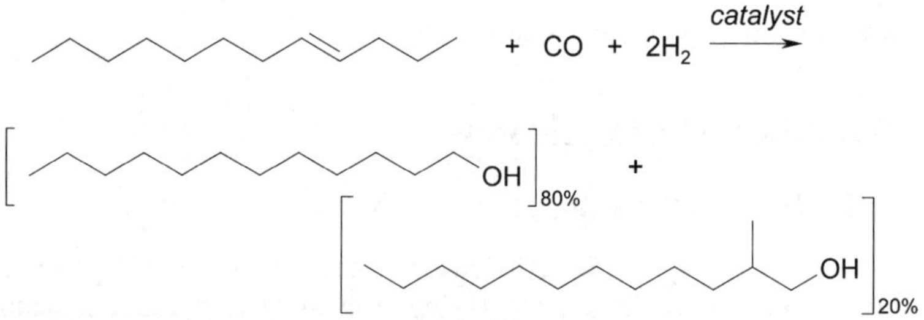

Figure 4.11   Modified OXO alcohol process showing the percentage production of the straight chain and branched products.

Figure 4.12 is a linear internal olefin. The top reaction shows the route to traditional NEODOL® alcohols, a Shell mixture of compounds, from an internal olefin. The lower reactions show how a linear internal olefin can be converted into a branched internal olefin. Standard modified OXO chemistry converts the branched internal olefin into a branched primary alcohol. Each of these structures depicts one of many possible isomers. The commercial product is >95% branched.

4.4.2.1   *OXO Alcohols Derived from Fischer–Tropsch α-Olefins*

This process begins with the production of syngas from either coal or natural gas. Syngas is then converted to a liquid hydrocarbon stream in the Fischer–Tropsch (F-T) process[79,80] (Figure 4.13).

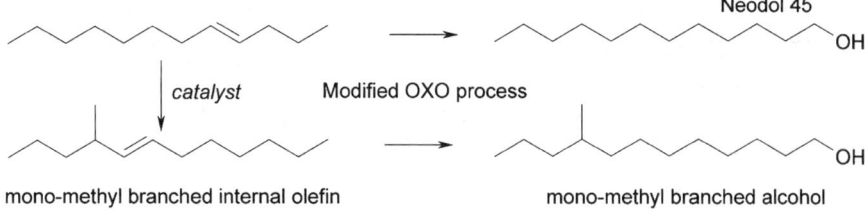

Figure 4.12 Modified OXO process for mid-range alcohols.

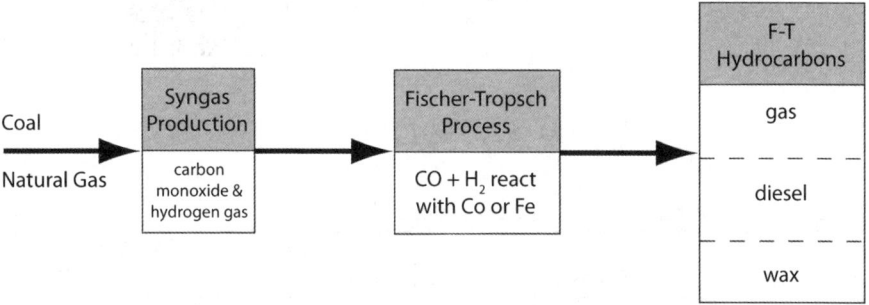

Figure 4.13 Fischer–Tropsch process.

This stream consists of both even and odd chain length hydrocarbons in a Shultz–Flory distribution; the principal components are α-olefins. The C_{11} and C_{12} hydrocarbons are separated by distillation. Hydroformylation, using CO and H_2, then serves to select the olefinic portion of that stream to make long chain aldehydes. Further hydrogenation and purification yield the F-T OXO 1213 alcohol. F-T alcohols are 50% branched, but randomly branched because of methyl branching in the precursor olefins. The process flow is summarised in Figure 4.14.[79,80]

4.5 Summary of Products

A comparison of the major detergent alcohols is given in Table 4.3.[79,80] An analysis of typical fatty alcohol mixtures provided for industrial and domestic formulations can be seen in Figure 4.15.

4.6 Detergent Formulations

Fatty alcohols are widely used in the manufacture of detergents; there are several types with (poly)ethoxylate or sulfate adjuncts imbuing the alcohol with increased water solubility. The most frequently used class of detergents with alcohol as the non-polar component are the alcohol ethoxylates (AEs); examples of typical structures are shown in Figure 4.16 and mixtures used in formulations in Table 4.4.

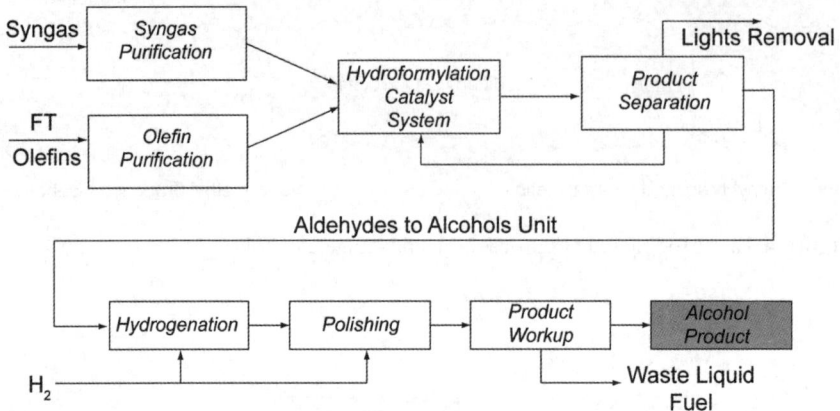

Figure 4.14   Fischer–Tropsch OXO alcohol process.

Table 4.3   Comparison of detergent alcohols.

Alcohol type	Raw material	Carbon chain distribution	Proportion linear (%)
Oleochemical	Coconut oil, palm kernel oil	C_{10}–C_{18} even only	100
Ziegler	Ethylene	C_2–C_{20} even only	98
Modified OXO	Ethylene	C_{12}–C_{15} even and odd	80
Regular OXO	n-Paraffin	C_{12}–C_{15} even and odd	50

There are two characteristics of AEs that determine their behaviour: chain length of the parent alcohol and number of ethoxylates attached to the alcohol. The principal chain lengths used in detergent manufacture are C_{12}–C_{16} and C_{18} n-alkanols.[81] The length of the alcohol reflects the synthetic pathway and solubility needs of the product. No data for commercial samples are readily available but some data from the analysis of sewage treatment plant (STP) effluents can be used to give an indication of the likely profile. Data from Western Europe, Canada and the USA are shown in Figure 4.17.[82]

The data shown in Figure 4.17 include natural fatty alcohols derived from a range of sources within the sewage treatment system as well as anthropogenically derived detergent alcohols.[83,84]

Detergent formulations use a series of ethoxylates up to ~20 carbons. Some representative data using two typical formulations can be seen in Figure 4.18. Influent material to a STP will contain AEs of this general formulation. This can be compared to measurements made in STP effluent from samples in different geographic regions (Figure 4.19). The $n=0$ sample is the free fatty alcohol and is substantially higher than the ethoxylates as it contains alcohols derived from non-detergent sources as well. The distinctive ethoxylate chain pattern (seen in Figure 4.18) of the commercial material has also been lost.

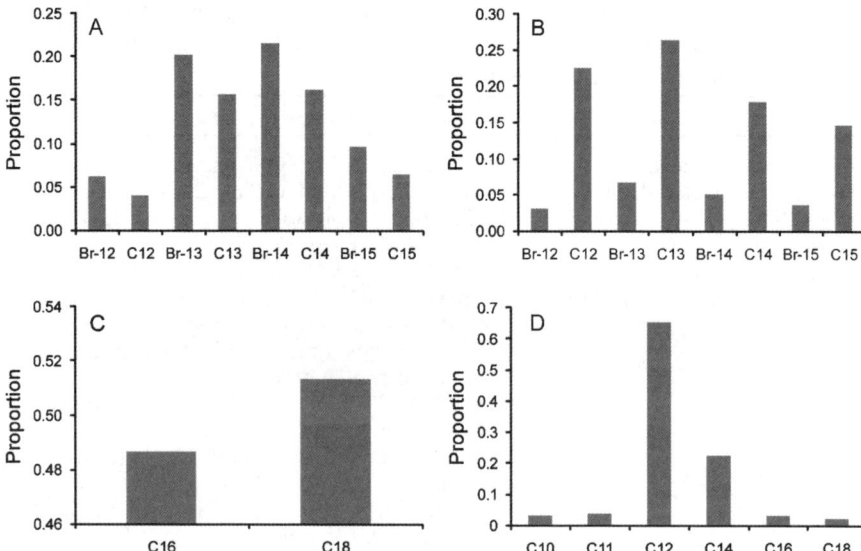

Figure 4.15 Fatty alcohol chain length profiles for four different raw materials used by industries in subsequent formulations. (A) A detergent range OXO alcohol with C_{12} to C_{15} produced by the dehydrogenation of linear paraffins and the subsequent hydroformylation of the resultant olefins. The alcohol is approximately 40–50% linear and 50–60% isomeric primary alcohols. (B) A detergent range alcohol with C_{12} to C_{15} carbon atoms. The alcohols are predominately linear, with approximately 20% 2-alkyl branching. The alcohols are derived from ethylene oligermisation, metathesis of some of the olefins to odd carbon lengths, and then the modified OXO reaction is applied to produce the alcohol. (C) A technical emulsifier-range fatty alcohol cut (C_{16}–C_{18}) derived from tropical vegetable oil, characterised by linear, even numbered carbon chains, obtained by catalysed high-pressure hydrogenation of the corresponding methyl esters. (D) A technical detergent-range fatty alcohol cut (C_{12}–C_{18}) derived from tropical vegetable oil, characterised by linear, even numbered carbon chains, obtained by catalysed high-pressure hydrogenation of the corresponding methyl esters.

Figure 4.16 Typical structures of some alcohol-based detergents; top: alcohol polyethoxylates where $n = 1$–20; bottom: alcohol sulfate, *e.g.* sodium dodecyl sulfate (SDS).

Table 4.4　Chemical Abstracts Service (CAS) registry numbers for blends of fatty alcohols used in detergent formulations with the principal chemical species present in each.

CAS	Chemical name	Composition
111-27-3	1-Hexanol	100% linear; >95% C_6 (range C_6–C_{10}); even
111-87-5	Octyl alcohol	100% linear; >90% C_8 (range C_6–C_{10}); even
85566-12-7	C_8–C_{10} alcohols	100% linear; >80% $C_{8/10}$, $C_6 \leq 5\%$, $C_{12} < 10\%$ (range C_6–C_{12}); even
112-30-1	1-Decanol	100% linear; >90% C_{10} (range C_8–C_{12}); even
68603-15-6	C_6–C_{12} alcohols	Generic: 5–100% linear; C_6–C_{12} alcohols (range C_6–C_{13}); even or even and odd Type A: 5–95% Linear; $\geq 95\%$ C_{11} (range C_9–C_{13}); even and odd Type B: >80% linear; >95% $C_{9/10/11}$ (range C_8–C_{12}); even and odd Type C: >80% linear; >95% $C_{7/8/9}$ (range C_6–C_{10}); even and odd Type D: 100% linear; $\geq 90\%$ $C_{8/10}$; <10% C_6 (range C_6–C_{12}); even
66455-17-2	C_9–C_{11} alcohols	>80% linear; >95% $C_{9/10/11}$ (range C_8–C_{12}); even and odd
85665-26-5	C_{10}–C_{12} alcohols	100% linear; >90% $C_{10/12}$, $\leq 5\%$ C_{14} (range C_8–C_{16}); even
112-42-5	Undecyl alcohol	>80% linear; >95% C_{11} (range C_9–C_{14}); even and odd
67762-25-8	C_{12}–C_{18} alcohols	Generic: 100% linear; >95% $C_{12/14/16/18}$ (range C_8–C_{20}); even Type A: 100% linear; >50% $C_{12/14}$; >10% $C_{16/18}$ (range C_8–C_{20}); even Type B: 100% linear; >10% $C_{12/14}$; >60% $C_{16/18}$ (range C_{12}–C_{20}); even
67762-41-8	C_{10}–C_{16} alcohols	Generic: 5–100% linear; C_{10}–C_{16} alcohols (range C_8–C_{18}); even or even and odd Type A: 100% linear; >80% $C_{10/12/14}$; <10% C_{16} (range C_8–C_{18}); even Type B: 5–50% linear; >95% $C_{12/13}$ (range C_{11}–C_{14}); even and odd Type C: 80–95% linear; >95% $C_{12/13}$ (range C_{11}–C_{15}); even and odd Type D: 40–50% linear; >95% $C_{12/13/14/15}$ (range C_{11}–C_{16}); even and odd
68855-56-1	C_{12}–C_{16} alcohols	Generic: 40–100% linear; C_{12}–C_{16} alcohols, >95% $C_{12/13/14/15}$ (range C_8–C_{18}); even or even and odd Type A: >40% linear; >95% $C_{12/13/14/15}$ (range C_{10}–C_{17}); even and odd Type B: 100% linear; >80% $C_{12/14}$, >20% C_{16} (range C_8–C_{18}); even Type C:. 100% linear; <10% C_{12}, >90% $C_{14/16}$ (range C_{10}–C_{18}); even
75782-86-4	C_{12}–C_{13} alcohols	>80% linear; >95% $C_{12/13}$ (range C_{11}–C_{15}); even and odd

Table 4.4 (*Continued*).

CAS	Chemical name	Composition
75782-87-5	C_{14}–C_{15} alcohols	>80% linear; >95% $C_{14/15}$ (range C_{12}–C_{17}); even and odd
80206-82-2	C_{12}–C_{14} alcohols	Generic: 100% linear; >95% $C_{12/14/16}$ (range C_6–C_{18}); even Type A: 100% linear; >90% $C_{12/14}$ ($C_{12} > C_{14}$), <10% C_{16} (range C_6–C_{18}); even Type B: 100% linear; >95% $C_{12/14}$ ($C_{12} < C_{14}$) (range C_8–C_{18}); even
63393-82-8	C_{12}–C_{15} alcohols	Generic: >40% linear; >95% $C_{12/13/14/15}$ (range C_{10}–C_{17}); even and odd Type A: >80% linear; >95% $C_{12/13/14/15}$ (range C_{10}–C_{17}); even and odd Type B: 40–50% linear; >95% $C_{12/13/14/15}$ (range C_{11}–C_{16}); even and odd
112-72-1	1-Tetradecanol	100% linear; >95% C_{14} (range C_{12}–C_{16}); even
68333-80-2	C_{14}–C_{16} alcohols	Generic: 5–95% linear; >95% $C_{12/13/14/15}$ (range C_{11}–C_{16}); even and odd Type A: 5–95% linear; >95% $C_{14/15}$ (range C_{12}–C_{17}); even and odd Type B: ≤5% linear; >95% $C_{12/13/14/15}$ (range C_{11}–C_{16}); even and odd
36653-82-4	1-Hexadecanol	100% linear; ≥ 95% C_{16} (range C_{14}–C_{18}); even
67762-27-0	C_{16}–C_{18} alcohols	100% linear (or unstated); <10% C_{14}, ≥ 90% $C_{16/18}$ (range C_{12}–C_{20}); even
67762-30-5	C_{14}–C_{18} alcohols	Generic: 100% linear (or unstated); >95% $C_{14/16/18}$ (range C_{10}–C_{20}); even Type A: 100% linear (or unstated); ≥ 95% $C_{16/18}$ (range C_{12}–C_{20}); even Type B: 100% linear (or unstated); >95% $C_{14/16/18}$ (range C_{10}–C_{20}); even
629-96-9	1-Eicosanol	>80% linear; ≥ 90% C_{20} (range C_{18}–C_{22}); even
97552-91-5	C_{18}–C_{22} alcohols	100% linear; >95% $C_{18/20/24}$ (range C_{16}–C_{24}); even
661-19-8	1-Docosanol	100% linear; >95% C_{22}; even
68002-94-8	C_{16}–C_{18} and C_{18} unsaturated	100% linear; >70% $C_{16/18}$, <10% C_{14}, including 40–90% C_{18} unsaturated (range C_{12}–C_{22}); even
68155-00-0	C_{14}–C_{18} and C_{16}–C_{18} unsaturated	Linearity unspecified; 5–50% $C_{16/18}$ saturated, 40–90% $C_{16/18}$ unsaturated (range C_{14}–C_{18}); even
112-70-9	1-Tridecanol	>80% linear; >90% C_{13}, <10% C_{12} (range C_{12}–C_{14}); even and odd
143-28-2	9-Octadecen-1-ol (9Z)	100% linear; >70% $C_{16/18}$, <10% C_{14}, including >70% C_{18} unsaturated (range C_{12}–C_{20}); even
629-76-5	1-Pentadecanol	>80% linear; >90% C_{15}, <10% C_{14} (range C_{14}–C_{15}); even and odd
68551-07-5	C_8–C_{18} alcohols	100% linear; 5–30% $C_{8/10}$, >60% $C_{12/14/16/18}$ (range C_8–C_{20}); even
90583-91-8	Tridecanol, branched and linear	5% linear; >95% C_{13}; odd

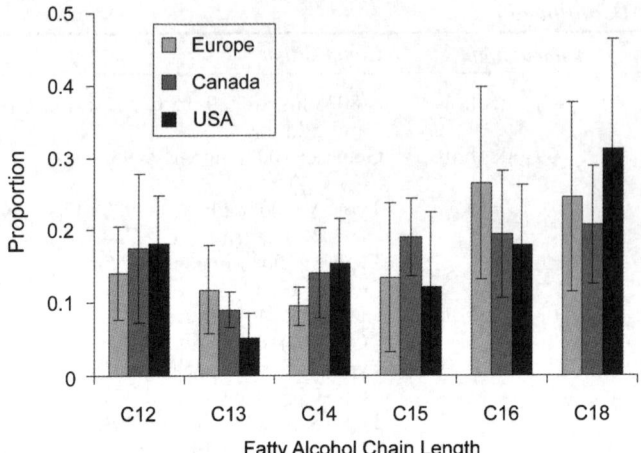

Figure 4.17 Mean free fatty alcohol chain length in STP materials (activated sludge, trickling bed filters or oxidation ditches) from Europe, Canada and the USA. The error bars are 1 standard deviation. The total fatty alcohol concentration range was $0.32–11.2\,\mu g\,l^{-1}$ for European samples, $0.29–14.2\,\mu g\,l^{-1}$ for Canadian samples and $0.13–2669\,\mu g\,l^{-1}$ for USA samples.

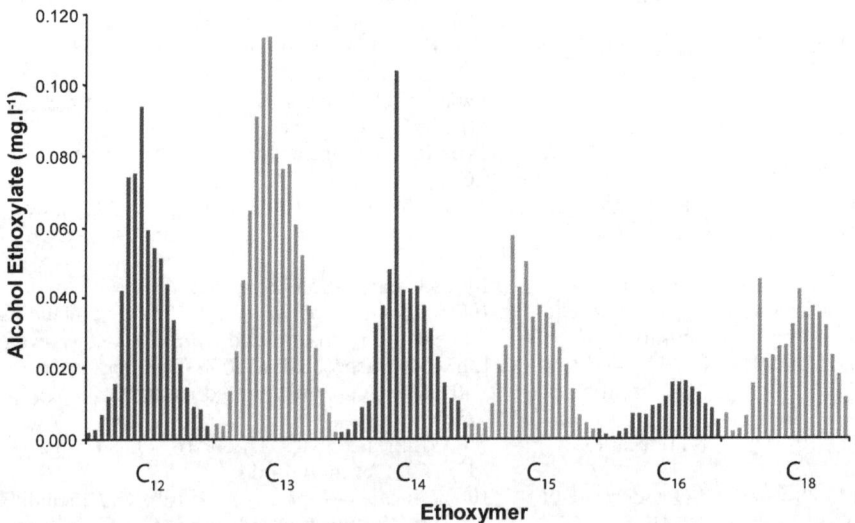

Figure 4.18 Ethoxylate chain length (0 to 18) in a mixture of commercial detergent formulations; this is considered typical of influent material to a STP. Data from Wind *et al.*[85]

The production of AEs is significant with close to 1 million tonnes produced annually worldwide.[81] The usage and production are centred in three regions: Japan, Western Europe and North America. The production in each, where available, is summarised below.

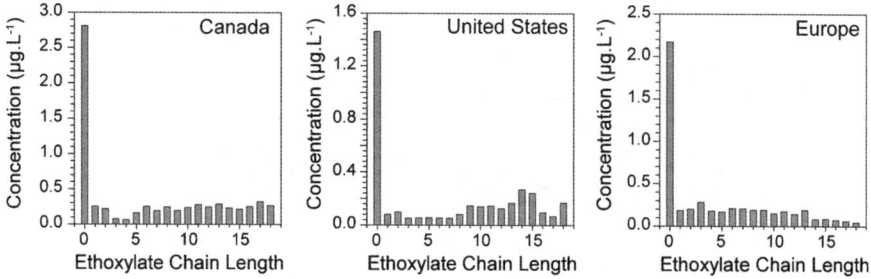

Figure 4.19 Distribution of ethoxylate chains by region in STP effluent. Alkyl chain lengths from C_{12} to C_{18} were summed per ethoxylate. Data from Eadsforth *et al.*[83] and Morrall *et al.*[84]

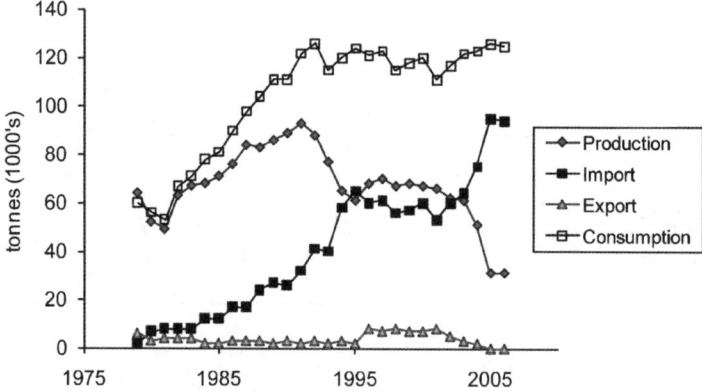

Figure 4.20 Production and usage of fatty alcohols as detergents in Japan. Data from Modler.[81]

4.6.1 Japan

The usage of fatty alcohols in detergents in Japan increased from 1979 to reach a peak in 1992 (Figure 4.20); since then, the consumption has stabilised at ~120 000 tonnes per year with approximately half of that quantity being imported. Production in Japan has declined since the peak in 1992, but has levelled off in recent years. In Japan, there has been a shift away from fatty alcohol production derived from plant oils (Figure 4.21) to those developed from petrochemical industries. A rapid decline in production began in 1992 but this has stabilised into a slower decline from the mid-1990s. The plant sources now include oils derived from palms and coconuts from Malaysia and the Philippines. From a fingerprinting viewpoint, the change in feedstock will result in different isotopic signatures, although with imports accounting for half of consumption, these may be blurred.

The consumption of these alcohols is principally to produce alcohol ethoxylate detergents (Table 4.5), although sulfates have been more important in the past environmental benefit of.[81]

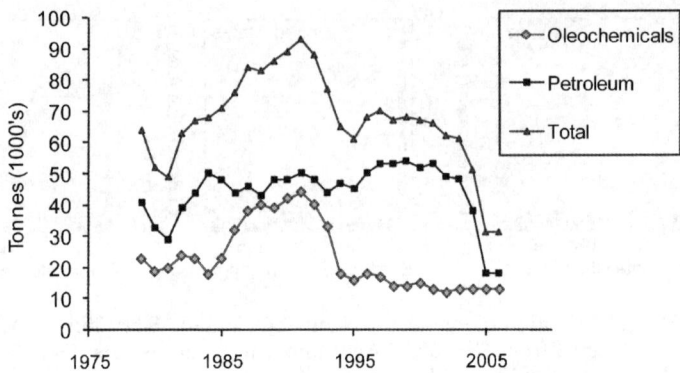

Figure 4.21 Production of fatty alcohols with Japan from natural and synthetic sources. Data from Modler.[81]

Table 4.5 The usage of fatty alcohols (in thousands of tonnes) in detergents in Japan with a forecast of the likely 2011 usage.

	1992	*1995*	*1998*	*2002*	*2006*	*2011*
Alcohol ethoxylates + AES	58	62	65	77	85	101
Alcohol sulfates	42	34	21	9	8	7
Other derivatives	13	16	18	21	23	27
Alcohols used as such	13	12	11	10	9	8

Data from Modler *et al.*[86]

Table 4.6 The usage of fatty alcohols (in thousands of tonnes) in detergents in Western Europe.

	1995	*1998*	*1999*	*2002*	*2006*
Alcohol ethoxylates	245	302	333	419	454
Alcohol sulfates	73	85	91	68	71
Polymethacrylate esters	27	29	30	31	34
Fatty nitrogen derivatives	11	16	20	23	26
Thiodipropionate esters	5	5	5	5	5
Other derivatives, alcohols used as such and C_{20}+ alcohols	64	72	75	81	82

Data from Modler *et al.*[86]

4.6.2 Western Europe

A similar story can be seen in Western Europe (Table 4.6); most alcohols are (and have been) used in the production of polyethoxylates. The growth in production has principally been led by the displacement of linear alkylbenzene sulfonate (LAS) surfactants with alcohol-based surfactants which have better compatibility with enzymes and higher efficacy in low- or non-phosphate powders. In Sweden and Denmark, environmental considerations have led to their usage over this period and a more favourable price *versus* performance relationship existed compared to linear alkylbenzenes (LABs).

4.6.3 North America

The production of fatty alcohols for use in detergents is focused in the USA, and of those used in Canada, most originate in the USA. The production by year and type of detergent manufactured can be seen in Table 4.7. There has been a large increase in the use of alcohol ethoxylates, although in recent years with a smaller increase in the use of alcohol sulfates. The end use of these alcohol-based detergents is principally in household and personal care products ($\sim 80\%$) with industrial applications amounting to $\sim 20\%$ of the total in 2003 (Figure 4.22), although this rose to 25% by 2006.[86] This latter use may increase further in the future as detergents based on nonylphenol polyethoxylates, which are known to have a poorer behaviour in the environment, are replaced by alcohol-based compounds.[81]

4.7 Future Fatty Alcohol Production

Future production of fatty alcohols may become less dependent on petroleum-based ingredients and rely more on natural production processes either *via* fatty acids that could be converted to fatty alcohols or direct production as alcohols. There is an opportunity to link the production of these compounds to carbon capture through CO_2 removal in the emissions from power stations and the like (see *National Geographic* magazine, October 2007 for an overview).[87]

There are many estimates of the amount of oil reserves remaining in the world, but 40 year's worth from 2008 based on current usage and dis covery rates has been mooted.[88] Increased demand for fuel and bio-based oils has led to a surge in the amount of agricultural land turned over to biofuel production. However, this has been at the expense of food production in some cases which may have a negative impact on our ability to feed the world's population in the future. In addition, the perception that biofuel production has a smaller environmental footprint than geologically derived petroleum is being seriously questioned.[89,90] There has already been some controversy regarding the planting of more palm trees to supply the current demand which has had deleterious effects on biodiversity even though they have high lipid outputs.[91,92]

There are potential crops that do not replace food crops, such as the poisonous scrub weed *Jatropha curcas*;[93] in this case, the plant produces high-quality seeds in environments that would otherwise not produce crops at all. It is being explored at present as a useful biofuel in countries such as India and China.

The price of oil and consequently products made from it will increase as the reserves are diminished.[94] Biomass, such as straw and wood by-products, may be metabolised by micro-organisms to form biofuels or other industrial raw materials.[95,96] Production of oils, lipids, fatty acids and high value biochemical

Table 4.7 The usage of fatty alcohols (in thousands of tonnes) in detergents in North America. There is an overall forecast growth to 2011 of 2.7% per annum.

	1980	1985	1988	1989	1990	1991	1992	1993	1994	1995	1996	1997	1998	2000	2002	2006
Alcohol ethoxylates	122	169	181	184	193	184	182	241	284	311	304	353	365	391	346	382
Alcohol sulfates	74	67	91	83	54	71	78	34	29	29	17	27	27	25	78	56
Polymethacrylate esters	27	17	15	14	14	13	13	13	16	11	12	11	12	13	15	17.5
Fatty nitrogen derivatives	9	7	7	7	6	6	9	11	11	11	9	9	9	10	10	12.5
Alkyl glyceryl ether sulfonates	12	8	5	5	5	5	5	5	6	7	8	7	7	7	7	8
Thiodipropionate esters	2	3	3	3	3	3	3	3	3	3	3	7	7	7	7	3.5
Other derivatives	5	6	7	7	7	7	7	15	18	16	16	14	14	14	17	11.5
Free alcohols, and all C_{20} + alcohols	22	23	24	24	24	24	24	25	29	28	27	27	28	27	31	30
Total	272	301	332	326	306	314	320	347	396	416	396	454	470	493	513	521.5

Data from Modler et al.[74,86]

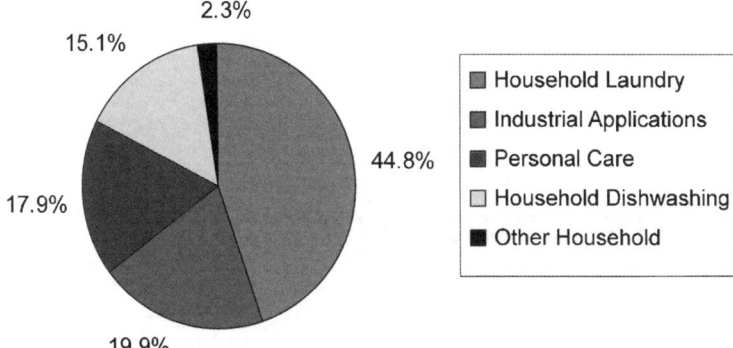

Figure 4.22 Usage of detergent fatty alcohols in North America. Data from Modler.[81]

products from microalgae was the subject of intense investigations in the past[97] and has re-emerged as economic pressure on petroleum-derived oil has increased in the past few years.[98] Bio-ethanol and fatty acids may be used as carbon sources for subsequent fatty alcohol production through the processes outlined in this chapter. Microbial production by *E. coli* also provides a worldwide opportunity to synthesise fatty acids.[99]

Summary

- Fatty alcohols of C_6 to C_{22} chain length are routinely used in a range of consumer products including cleaning liquids and personal care formulations. Up to 25% of the product may be fatty alcohols.
- These fatty alcohols may be derived from oil (petrochemical) or biological (oleochemical) sources. The former materials can be synthesised from ethene or paraffin after a range of processes. The latter source usually involves hydrogenation of fatty acids or fatty acid methyl esters from palm products.
- Fatty alcohols derived from natural materials tend to be even carbon chains with no branches while petorchemically derived materials may also include substantial amounts of branching and include odd carbon chain lenghts.
- Fatty alcohols are often used as their sulfates or polyethoxylates with over 1 million tonnes used annually around the world.

CHAPTER 5
Environmental Transformations

Once in the environment and sewage treatment system, what happens to the fatty alcohols? There are contrasts between the short and long chain moieties which have very different degradation rates.

All organic matter is potentially degradable in the environment and may be utilised in the short term by microbial action or over longer periods through geochemical processes leading to the formation of kerogen or oil. The debris from cell lysis, leaf fall, feeding, faecal matter and organism death generally are accumulated in the soils or surface sediments. Some material may be degraded during its settling through the water column[100] and the material that reaches the sediments is likely to be altered considerably compared to the original source material. There will also be a transfer of terrestrial materials to the sea by either wind blow[37] or more usually through river flow and soil wash-off.

Lipids are substantially more resistant to degradation than carbohydrates or proteins and often survive better to reach the sediments more intact than other organic matter. In general, small molecules will be more rapidly degraded than large ones and aliphatic compounds are degraded more quickly than aromatic structures. This means that waxes are likely to reach the sediments reflecting the original source material and free fatty alcohols (and acids) are less likely to do so.

5.1 Metabolism of Fatty Alcohols

Fatty alcohols are metabolised as part of a system that operates on alkanes and fatty acids as well. The general scheme is shown in Figure 5.1. In the oxidation process, alkanes may be converted to alcohols and subsequently fatty acids. These then enter the β-oxidation pathway to yield a series of acetyl-CoA products (and one proprionyl-CoA in the case of odd chain compounds) and several molecules of ATP.[16,101] This process happens within the cell and the

Fatty Alcohols: Anthropogenic and Natural Occurrence in the Environment
By Stephen M Mudge, Scott E Belanger, and Allen M Nielsen
© Copyright 2008 ERASM (the joint surfactant environmental research platform of AISE and CESIO) and SDA
Published by the Royal Society of Chemistry, www.rsc.org

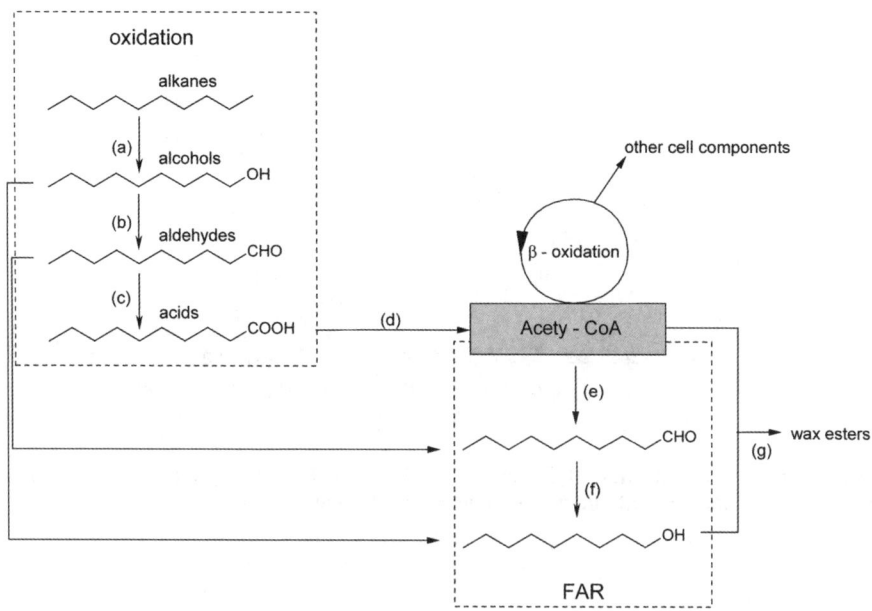

Figure 5.1 Schematic process for the metabolic degradation of fatty alcohols in *Acinetobacter* spp. The oxidation component is degradative while the FAR step builds new compounds. (a) Alkane monooxygenase, (b) alcohol dehydrogenase, (c) aldehyde dehydrogenase, (d) acyl-CoA synthetase, (e) acyl-CoA reductase, (f) aldehyde reductase (alcohol dehydrogenase) and (g) acyl-CoA:alcohol transferase. Redrawn from Ishige *et al.*[102]

products of this oxidation pathway are used to provide energy, water and compounds suitable for other metabolic processes.

Environmental transformations of alkanes and waxes may be mediated by bacteria (*e.g.* hydrocarbon degradation) and these reactions yield alcohol intermediates. It is worth noting at this point that naturally occurring bacteria are able to degrade waxes.[103] The bacterial (*Pseudomonas oleovorans*) alkane hydroxylase system[102] that is responsible for the total oxidation of an *n*-alkane to *n*-alcohol $(RCH_3 + NADH + H^+ + O_2 \rightarrow RCH_2OH + NAD^+ + H_2O)$ consists of three components: alkane hydroxylase (AlkB), rubredoxin (AlkG) and rubredoxin reductase (AlkT). AlkB is a non-heme iron integral membrane protein that catalyses the hydroxylation reaction. AlkG transfers electrons from the NADH-dependent flavoprotein rubredoxin reductase to AlkB. The resultant alcohol is oxidised to 1-alkanoate by a membrane-bound alcohol dehydrogenase (AlkJ) and cytosolic aldehyde dehydrogenase (AlkH). 1-Alkanoate is incorporated through β-oxidation *via* the acyl-CoA synthetase (AlkK) reaction.

This is the case for relatively small compounds with chain lengths between C_5 and C_{12}.[102] Other Gram-negative *n*-alkane degraders belonging to *Acinetobacter* grow on longer chain *n*-alkanes. Although the reactions for the longer chain

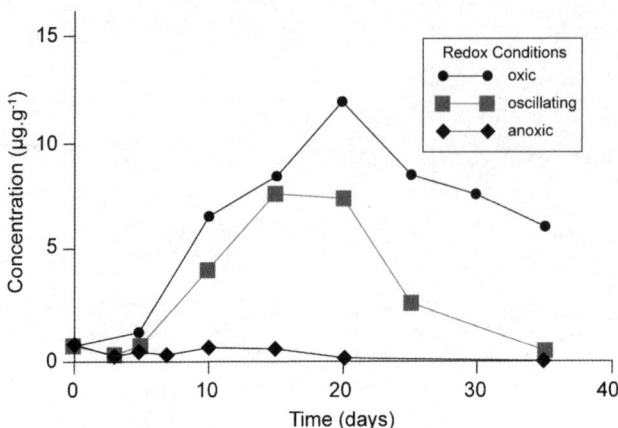

Figure 5.2 Concentration of C_{16} and C16:1 fatty alcohols during oxic, oscillating and anoxic incubations of sediment with the microalga *Nannochloropsis salina*.[104]

alkanes C_{12}–C_{18} are principally the same as those of *Pseudomonas oleovorans*, the organisation of the genes is different.[102] *Acinetobacter* sp. strain M-1 is characterised by its ability to use much longer chain *n*-alkanes (C_{20}–C_{44}) and can degrade *n*-alkanes up to C_{60} when grown on a paraffin wax mixture.[102]

Work by Caradec *et al.*[104] on the degradation of fatty acids identified the production of free saturated and mono-unsaturated C_{16} and C_{18} fatty alcohols during anoxic and alternating oxic/anoxic incubations of algal material and natural sediments. The production of fatty alcohols coincided with a high degree of triacylglycerol hydrolysis. This supports a precursor–product relationship between fatty acids esterified to triacylglycerol and the alcohols produced. The smaller production of C_{16} alcohols observed under anoxic conditions (Figure 5.2) might reflect a lower efficiency of anaerobic bacteria for mineralising these precursors. For the oscillating conditions, it is possible that alcohols were produced under anoxia and consumed during oxic periods. This suggests that aerobic and anaerobic bacteria in the sediments used different assimilation pathways.

Relatively little information exists on the extracellular production of free fatty alcohols, but some work on the fungus *Botrytis cinerea*[105,106] showed that some were indeed produced with a profile compatible with direct loss after FAS and FAR (57% C_{16} and 43% C_{18} and no longer or unsaturated compounds). There was a wider range of compounds present in waxes with the majority being C_{20} or C_{28}.

Some experimental work[106,107] has shown that exogenous fatty alcohols may be taken up by bacteria and in this case incorporated into bacteriochlorophyll-c. Usually, mid-chain length compounds were utilised (C_{16}–C_{18}), although it was demonstrated that short chain (C_{10}–C_{12}) and long chain (C_{20}) compounds were also utilised.

5.2 Natural Degradation

The degradation of short chain compounds occurs at a greater rate than that of long chain compounds. In their study of fatty acids in the marine environment, Haddad *et al.*[108] found the concentration of 16:0 decreased from 185.1 μg g^{-1} dry weight (DW) at the surface to 4.9 μg g^{-1} DW at 250 cm depth in a core from Cape Lookout Bight, NC, USA. In contrast, the 28:0 fatty acid only decreased from 16.3 to 11.2 μg g^{-1} DW over the same depth. This indicates the general pattern of degradation for chains of alkanes, alcohols and acids. In Figure 5.3, the relatively high concentration of short chain fatty acids at the surface ($\sim$350 μg g^{-1}) reduced to approximately 25% of this value within the top 25 cm. The corresponding long chain fatty acid concentration increased in absolute terms over the same depth interval; the change in these concentrations down the whole 250 cm was negligible.

5.2.1 Short Chain Moieties

Models of the relative fatty acid degradation rate have been constructed by Haddad *et al.*[108] In general terms, there was an order of magnitude less degradation of the long chain ($>C_{20}$) compounds compared to the short chain (C_{18} and below) compounds. There was also a 25% increase in the degradation rate for the even chain (*e.g.* C_{14} and C_{16}) compounds compared to the odd chain equivalents (C_{15} and C_{17}) (Figure 5.4).

In sediment samples from a core collected in a Scottish sea loch (essentially, a fjord; see Figure 5.5 for the location), the concentration of the short chain fatty

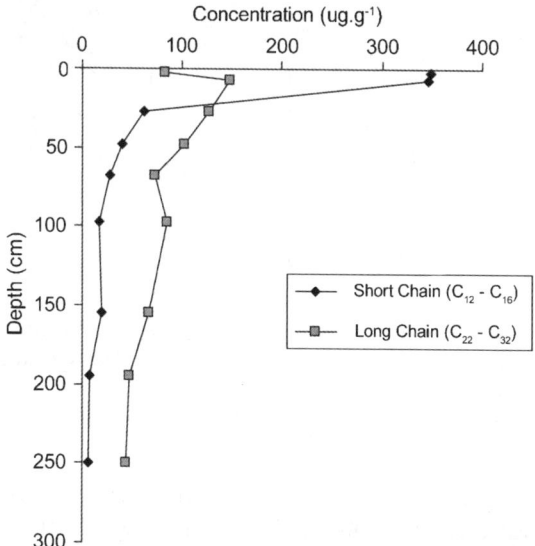

Figure 5.3 Fatty acid concentrations in a core. Data redrawn from Haddad *et al.*[108]

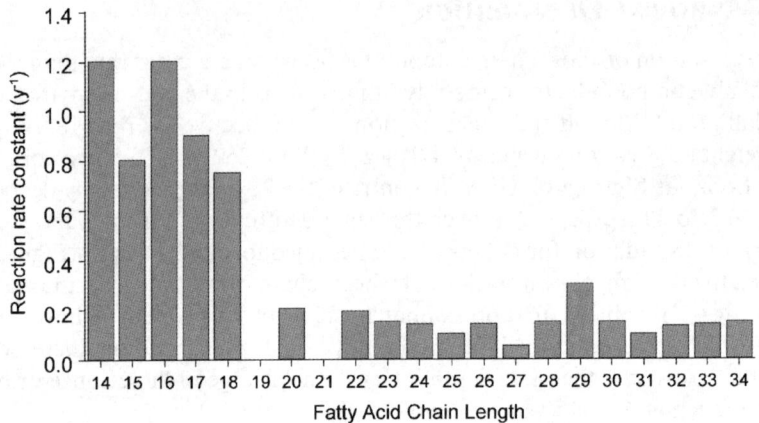

Figure 5.4 Degradation rates for fatty acids modelled by Haddad *et al.*[108]

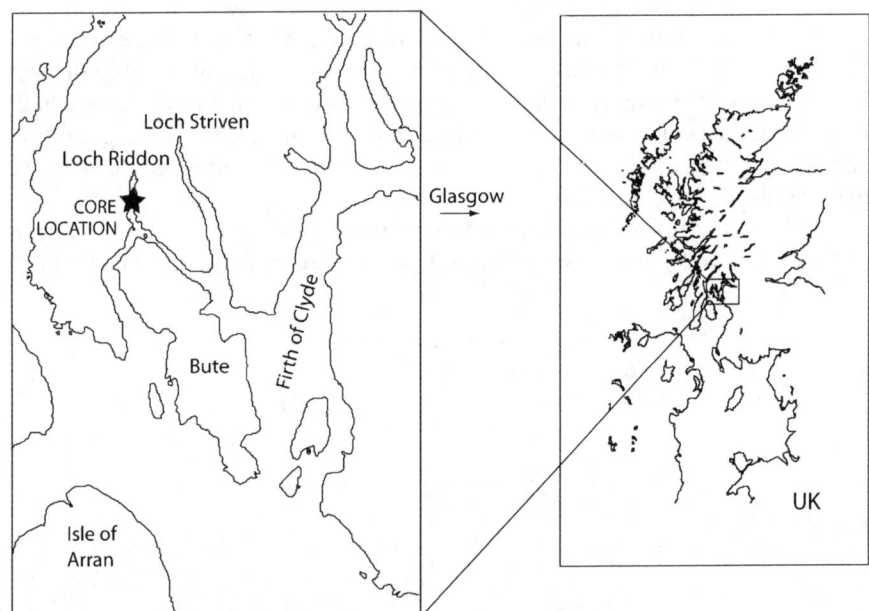

Figure 5.5 Location map for a 1.5 m core analysed for fatty alcohols and other lipid biomarkers taken from Loch Riddon, UK. The sills in sea lochs tend to trap both terrestrial matter that runs off from the land as well as fine-grained sediments that enter from the sea and then settle out.

alcohols generally decreased with depth (Figure 5.6a), especially in the top 20 cm, while the long chain compounds increased with depth (Figure 5.6b). The net effect of this can be seen in the short/long chain fatty alcohol ratio (Figure 5.6c). In the surface sediments there are greater concentrations of short

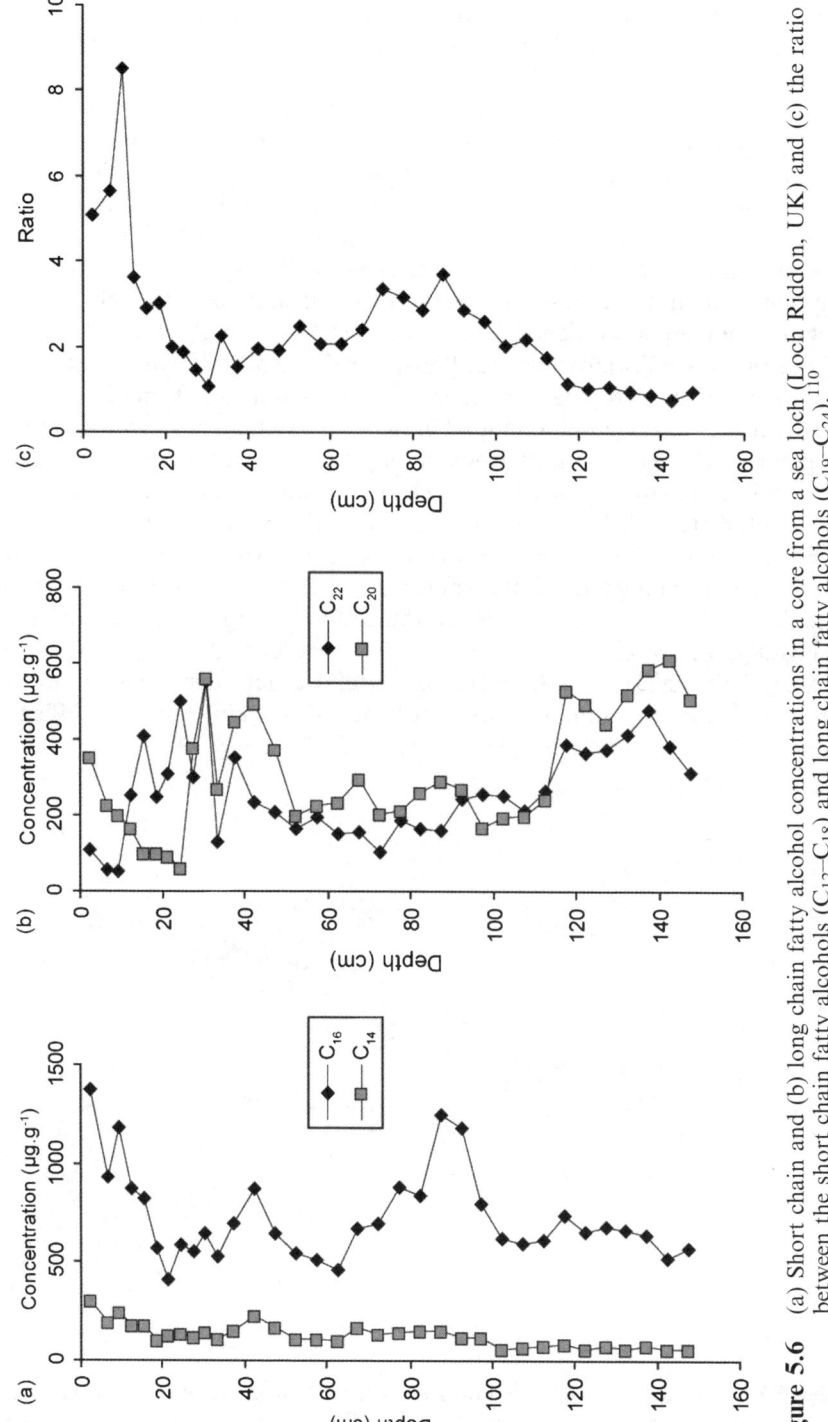

Figure 5.6 (a) Short chain and (b) long chain fatty alcohol concentrations in a core from a sea loch (Loch Riddon, UK) and (c) the ratio between the short chain fatty alcohols (C_{12}–C_{18}) and long chain fatty alcohols (C_{19}–C_{24}).[110]

chain, marine derived compounds often with a subsurface maximum. This is a common feature of several sediment cores[109,110] and may be due to *in situ* biosynthesis by bacteria utilising the depositing organic matter.

At the deeper depths, the concentration of the longer chain compounds is greater than the short chain ones and a ratio of less than one can be measured. This may reflect two distinct processes: *in situ* degradation of the short chain compounds or a change in organic matter source from marine at the surface (recent past) to terrestrial at depth (past 500 years). The difficulty in separating these two different processes will be considered in Chapter 9.

In some locations, terrestrial inputs are small and the short chain fatty alcohols dominate at all depths within the sediment core. An example of this can be seen in Ria Formosa lagoon, Portugal (Figure 5.7).[111] This is one of the largest lagoons in Europe and receives little terrestrial runoff for most of the year; when it does rain in November–February, the water tends to flush out any fine-grain suspended materials from the lagoon.[112,113] Therefore, the settled sediment is dominated by marine markers which can be seen in the short chain fatty alcohol (C_{12}–C_{18})/long chain (C_{19}–C_{24}) ratio (Figure 5.8). The ratio is considerably greater than the data for the sea loch environment which will be receiving and trapping terrestrial organic matter. Therefore, the absence of long chain fatty alcohols indicates a marine source for the organic matter and tells the investigator significant information about sources and their deposition in the area. This makes the fatty alcohols a useful group of biomarkers in the marine environment, although the sterols are also useful but in a different context.[56] These aspects are explored further in Chapter 8.

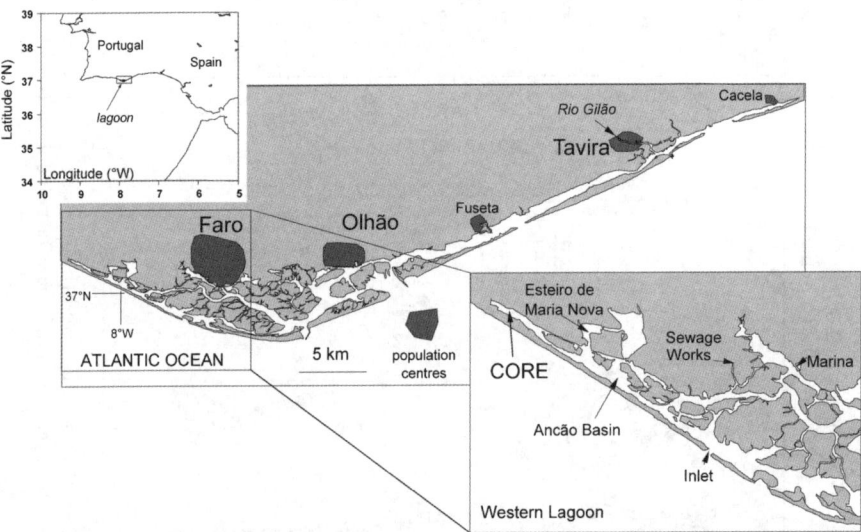

Figure 5.7   Ria Formosa lagoon in Portugal. Although there are rivers and other sites of terrestrial runoff, the region is dominated by marine-derived fatty alcohols as rainfall is principally confined to the winter months.

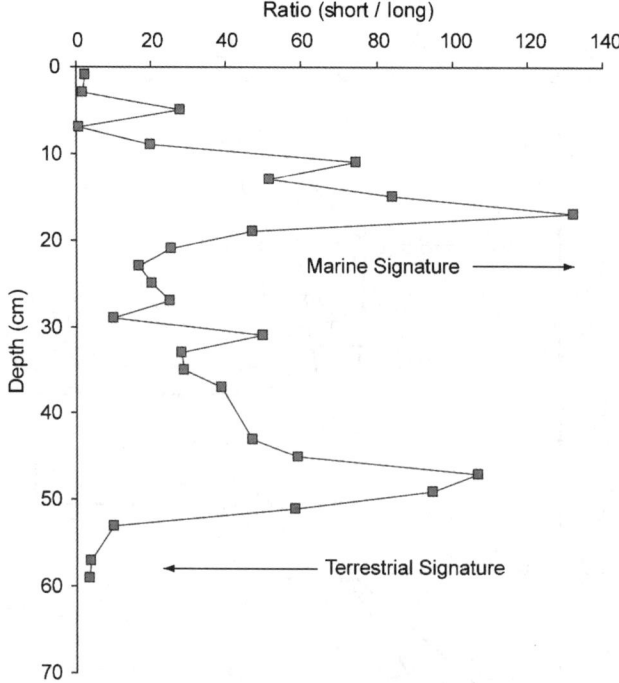

Figure 5.8 Short chain fatty alcohols (C_{12}–C_{18})/long chain fatty alcohols (C_{19}–C_{24}) ratio down a core from Ria Formosa lagoon. After Unsworth.[111]

5.2.2 Long Chain Moieties

Work on the settling and settled organic matter from the equatorial Pacific Ocean[114] has indicated that most biogenic compounds are degraded in the water column (Figure 5.9). Compounds that occurred at maximal abundance at 10–12 cm in the sediments were classed as Group IV; these were resistant to microbial degradation. Chemicals with this behaviour include three separate homologous suites of high molecular weight, straight chain fatty acids, fatty alcohols and alkanes. Compounds comprising these series exhibit carbon-number predominance patterns characteristic of cuticles from vascular land plants and likely were transported to the central Pacific by wind.[115,116]

Therefore, the longer chain fatty alcohols and acids reach the sediments and represent refractory constituents of water column particles that become magnified at depth by selective preservation as >99% of the surface-produced organic carbon was respired.

The degradation pathway for fatty alcohols follows the general scheme advanced in Figure 5.1: alkanes and alcohols are converted to fatty acids which enter the β-oxidation pathway.[101] The fate of the acetyl-CoA subunits cleaved off during this process is usually to end up as carbon dioxide. The short term diagenesis of fatty acids in marine sediments, the ultimate fate of most organic carbon, indicated the following reactivity relationships:[108] unsaturated fatty

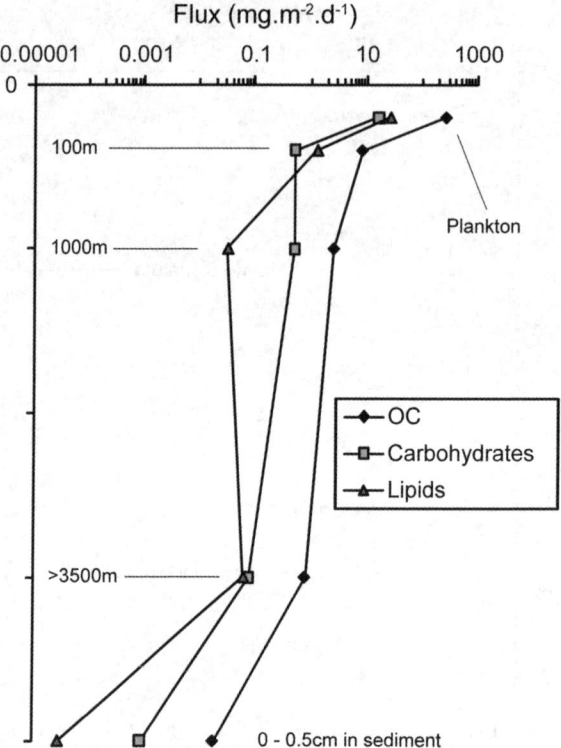

Figure 5.9   Data[114] showing the rate at which different organic matter classes degrade with depth through the water column and then into surface sediments in the open ocean at the equator.

acids > branched fatty acids > saturated fatty acids. Also detected were differences within the saturated fatty acid fraction such that medium chain length compounds (C_{14}–C_{19}) were degraded at rates 6–7 times faster than long chain length compounds (C_{20}–C_{34}). Results of kinetic modelling indicated that no simple relationship exists between remineralisation rates and molecular weight (or carbon chain length) and it was suggested that the preferential preservation of terrestrially derived long chain length fatty acids results from their inclusion into microbially inaccessible matrices.[108]

5.3   Degradation Rate Constants

Haddad *et al.*[108] calculated apparent degradation rate constants for *n*-alkanols and phytol by assuming that they are degraded by first order kinetics and are at steady state using the following equation:

$$\ln C = \ln C_0 - k\left(\frac{z}{s}\right)$$

where C = the alcohol concentration at depth; C_0 = the alcohol concentration at $z = 0$ (the surface); k = the apparent rate constant (y^{-1}); z = core depth (cm); s = sedimentation rate (cm y^{-1}).

The apparent rate constant can be estimated from the linear regression of $\log_e$ alcohol concentrations *versus* z/s. Results from the continental slope off Taiwan (354 m water depth and 0.33 cm y^{-1} sedimentation rate) are summarised in Table 5.1.

The values in Table 5.1 are similar to other published rates from Sun and Wakeham[118] who measured values between 0.024 and 0.070 y^{-1} in three locations. However, re-analysis of data from Loch Riddon[110] produces a slower degradation rate. Plotting of the $\log_e$ concentrations of the C_{14} and C_{16} fatty alcohols against k/s (from the equation above) can be seen in Figure 5.10. The sedimentation rate, s, was calculated from the position in the sediment core of the increased polyaromatic hydrocarbon (PAH) concentrations derived from the increased burning of coal and coke from 1750 onwards. In this core, that position was at 52.5 cm depth and so a sedimentation rate of 0.21 cm y^{-1} was used.

Table 5.1 Degradation rate constants (y^{-1}) for fatty alcohols in marine sediments.

	Extractable[a]	Bound[b]
Phytol	0.015	0.011
n-Alkanols	0.010	0.007

[a]Those that can be removed from the sediment without a saponification step.
[b]Those that need a saponification step. After Jeng *et al.*[117]

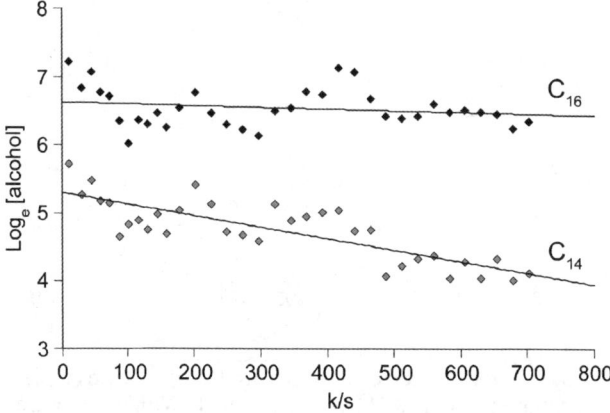

Figure 5.10 $\log_e$ of the C_{14} and C_{16} fatty alcohol concentrations from a Scottish sea loch. A sedimentation rate of 0.21 cm y^{-1} was calculated from the PAH profile based on the beginning of the industrial revolution *ca* 1750.

The calculated slope for the C_{14} was $0.002\,y^{-1}$, less than previously reported values for bound fatty alcohols shown in Table 5.1.[117] This may be for several reasons: incorrect sedimentation rates, although other analyses have confirmed the rate; better preservation in this particular site due to low bacterial activity; or altered input fluxes through time. The latter appears to be most likely as other biomarker signatures change with time due to increased anthropogenic organic matter deposition after the initial industrialisation period. The degradation rate of the C_{16} was almost an order of magnitude less than that of the C_{14}. This may be due to its increased chain length[117] or *in situ* production by biota. Shorter chain alcohols were only present in the top few centimetres of the sediment core (*e.g.* C_{12} to 12.5 cm) and have, therefore, degraded much quicker than the rate reported here.

5.3.1 Phytol Degradation

The degradation of phytol in the Loch Riddon marine core can be seen in Figure 5.11. The calculated degradation rate in this case was 0.005, which was half of the values of Jeng *et al.*[117] Again, the values could be compromised by increased primary productivity in the more recent past, fuelled by anthropogenically derived nutrients introduced as Glasgow expanded in the post-industrial revolution period.

The rate of phytol degradation compared to straight chain, saturated alcohols is greater; Jeng *et al.*[117] attributed this to the unsaturated nature of the molecule and compared their results with those of Sun *et al.*[119] who studied fatty acids in oxic and anoxic marine sediments.

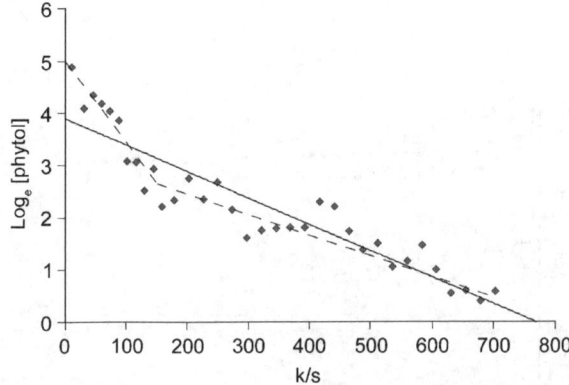

Figure 5.11 Log$_e$ of the phytol concentrations from a Scottish sea loch. A sedimentation rate of $0.21\,cm\,y^{-1}$ was used. Within the overall degradation rate shown by the solid line, there may be two different rates (dashed lines) with faster rates seen near the surface and slower rates at deeper depths.

5.4 Effect of Chemical Associations on Transformation Rates

5.4.1 "Natural" Fatty Alcohols in STPs

In a study of the potential origin of organic matter on Blackpool Beach, UK, analyses of fatty alcohols were made of the influent and effluent of the STP serving the area.[120] The profile of these chemicals can be seen in Figure 5.12; the concentration of the fatty alcohols in the influent material (liquid/liquid extraction of total after addition of KOH) is significantly greater than that in the effluent indicating removal/transformation of the materials during treatment.

The final effluent total fatty alcohol concentration is 0.16% of the influent total fatty alcohol concentration. In both cases, the profile is dominated by odd carbon chain alcohols: C_{19} in the influent and C_{17} in the effluent. This indicates the presence of high bacterial biomass as would be expected in these materials. The influent also has a range of long chain alcohols up to C_{28} indicative of

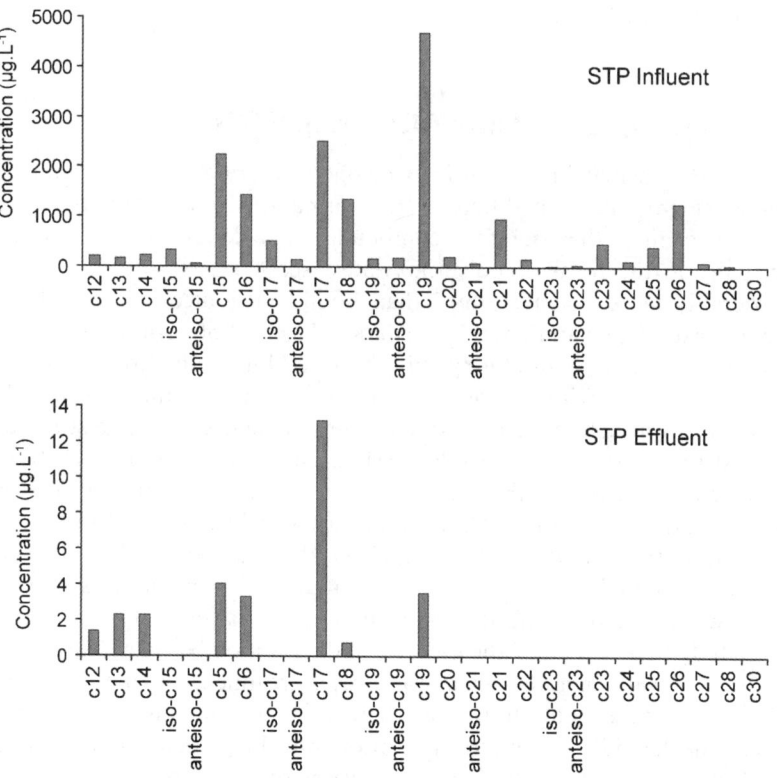

Figure 5.12 Fatty alcohols in the influent and effluent of the STP serving Blackpool, UK.

terrestrial plant matter. These would be associated with waste food material and terrestrial runoff entering the wastewater collection system. These materials are absent from the effluent, with the longest chain alcohol detected being C_{19}. The proportion of the short chain C_{12}–C_{14} compounds goes up due to the metabolism of the longer moieties or removal of these compounds with settling particulate matter. The lack of branched chain fatty alcohols is of interest as these are normally associated with bacteria.[50] Their absence may indicate a low biomass of the particular organisms that produce these *iso* and *anteiso* branches, that they are more readily metabolised or simply that their presence is below the limit of detection.

This pattern of degradation does not follow that seen in the environment; in natural sediments, the longer chain waxy materials are preserved and can be seen to considerable depths in cores. In STPs, the long chain materials are removed and the proportion of short chain ones increases. The origin of the C_{12}–C_{14} compounds in this system cannot be determined from these analyses alone as they may have originated from another, non-biological source, such as detergents, although these also degrade in STPs. Further work using compound-specific $\delta^{13}C$ may elucidate the origin of the short chain compounds in the effluent (see Chapter 8).

5.4.2   Anthropogenic Fatty Alcohols in STPs

The different chemical forms of fatty alcohols entering the wastewater system from detergents undergo different transformations during treatment, which ultimately changes the fate of the compounds. The detergent fatty alcohols are principally in the form of polyethoxylates (see Chapter 4) and as such are considerably more "bioavailable" than bound fatty alcohols, usually in the form of waxes. Itrich and Federle[121] studied the effect of both the alcohol chain length and the number of ethoxylate groups used in the hydrophilic section of the molecule. As with previous studies,[122–124] they utilised radiolabelled homologues and determined the fate of the radioactive ^{14}C. The study used a poisoned (control) activated sludge system and in this system >98% of the introduced radiolabelled ethoxylates were present as the parent molecule at the end of the experiment. In contrast, in the biologically active system no ^{14}C was recovered as the parent ethoxylate and 68.7% was measured in the trapped CO_2. Of the remainder, 7–14% of the initial ^{14}C from the α-carbon of the alcohol was associated with the solids, potentially as free fatty alcohols.[121]

In a study by Fisk *et al.*,[5] the theoretical distribution between phases and the fraction that was degraded in a wastewater treatment plant was calculated (Table 5.2). These data show that there is a big difference in partitioning between the degradation and incorporation into sludge that occurs around C_{11} and C_{12}. This is principally due to the water solubility (or lack of it) in the longer chain compounds. Bacteria are able to degrade these compounds, although the rates depend on the species present, their acclimation and the chain length and chemical association of the fatty alcohols. Fisk *et al.*[5] also

Table 5.2 Distribution of individual fatty alcohols in a wastewater treatment plant (WWTP) predicted by SimpleTreat model (using degradation rate).

	Fate in WWTP			
n-Alcohol chain length	*Fraction to air*	*Fraction to water*	*Fraction to sludge*	*Fraction degraded*
6	0.004	0.001	0.002	0.993
7	0.005	0.001	0.003	0.991
8	0.004	0.001	0.005	0.990
9	0.006	0.0014	0.0083	0.984
10	0.0076	0.0014	0.017	0.974
11	0.013	0.0014	0.019	0.967
12	0.0048	0.0038	0.47	0.521
13	0.00755	0.00417	0.498	0.490
14	0.00277	0.0077	0.597	0.393
15	0.00136	0.012	0.648	0.339
16	0.00172	0.0152	0.671	0.312
18	0.00163	0.0273	0.729	0.242
20	0.00014	0.045	0.796	0.159
22	0.000094	0.045	0.796	0.159

Data from Fisk *et al.*[5]

report the anaerobic degradation of fatty alcohols, although there are a limited number of such studies.

5.4.3 Sources of Fatty Alcohols in the Environment

One of the key unknowns with regards to the source of fatty alcohols in the aquatic environment is what proportion of them arise through the sewage treatment system. The origin of C_{12}–C_{16} in sewage effluent may be from:

(i) detergents produced using either oleochemical or petrochemical raw materials;
(ii) natural source materials such as terrestrial plant matter and human wastes;
(iii) degradation of longer chain compounds such as those found in terrestrial plants but reduced in chain length due to bacterial action;
(iv) new synthesis by the STP biota; and
(v) some mix of all these processes.

Determining the relative importance of each of these sources could be addressed by measuring the compound-specific $\delta^{13}C$ values for the alcohols in the environment, influent, effluent and source materials (see Chapter 8). Initial investigations suggest that human wastes are the most important source in sewage influent, although detergents based on oleochemical raw materials may also contribute but they could not be separated unambiguously from non-detergent sources in these experiments. What is also known, however, is that the stable isotope signature in marine sediments near the outfall of a STP is different to the influent values and so these materials probably make only a small contribution compared to the natural marine faunal production.

Alternatively, fatty alcohols may be produced within the sediments of the marine environment. These may result from *de novo* synthesis but also from the degradation of organic matter transferred to the system. Higher concentrations are usually measured at the surface of a core compared to those at depth: are these being produced *in situ* by degradation of long chain compounds, or is sediment compaction increasing the apparent concentration or bacterial biomass? While some data are available on the degradation of fatty acids,[108] little is known about fatty alcohol production or degradation at high resolution.

As well as synthesis, the rate of production of free fatty alcohols from waxes in natural conditions is largely unknown. Most alcohols in the environment are wax esters and little information is available on their rate of conversion to free alcohols. This has a direct relevance to the measured concentration when using a non-saponifying method such as that used for polyethoxylate quantification (see Chapter 6).

The mechanism for the production of free fatty alcohols from fatty acids under natural conditions exists but little is known about rates. Fatty acids exist in the environment both esterified as waxes but also linked to other compounds such as glycerol, but little is known about the exogenous production of free alcohols from these acids. There may be continued synthesis/degradation of fatty alcohols from fatty acid precursors exogenously in sediments. These rates are not quantified.

Summary

- Fatty alcohols and fatty acids are available for degradation in the environment although the form of these compounds (bound or free) will directly affect the rate at which they are metabolised.
- Alcohols may be converted to acids and become part of the β-oxidation pathway and become new biomass or CO_2.
- In general, the short chain compounds are more readily degraded in the environment than the longer chain waxy materials. Rates vary between 0.002 and 0.02 y^{-1} across a range of sites. However, in a sewage treatment plant, the long chain compounds are lower in the effluent compared to the influent, although there is production of new odd chain, bacterially derived compounds instead.

CHAPTER 6
Analytical Methods

How do we measure these compounds? Are there major differences in biases between methods?

6.1 Overview of Methods

There are two major approaches to the analysis of fatty alcohols in the environment. In one case, free fatty alcohols and polyethoxylates are measured (principally) by a liquid chromatography (LC) technique and in the other ester linked compounds are saponified and analysed by gas chromatography (GC). The former method seems to have become the standard for the detergent industry while environmental analytical laboratories are using the latter. This may lead to an issue of "context"; results generated by the former method will not include the ester linked compounds which form the bulk of the natural fatty alcohols. Therefore, when the latter method is applied no measure is made of the relative proportion of fatty alcohols in the environment derived from each source.

6.2 Methods for Analysis of Free Fatty Alcohols (and Ethoxylates)

The basis for this method is the derivatisation of the terminal –OH group with 2-fluoro-N-methylpyridinium p-toluenesulfonate (Pyr +), giving the molecule a net cationic charge.[125] The schematic of the reaction can be seen in Figure 6.1. Since all terminal –OH groups are derivatised, all species including the free fatty alcohols and those with only one ethoxylate can be effectively detected by electrospray mass spectrometry (MS).

The typical ethoxylates used in detergents have aliphatic alcohols with chain lengths between C_{12} and C_{18} (except C_{17}) with ethoxylate (EO) hydrophilic

Fatty Alcohols: Anthropogenic and Natural Occurrence in the Environment
By Stephen M Mudge, Scott E Belanger, and Allen M Nielsen
© Copyright 2008 ERASM (the joint surfactant environmental research platform of AISE and CESIO) and SDA
Published by the Royal Society of Chemistry, www.rsc.org

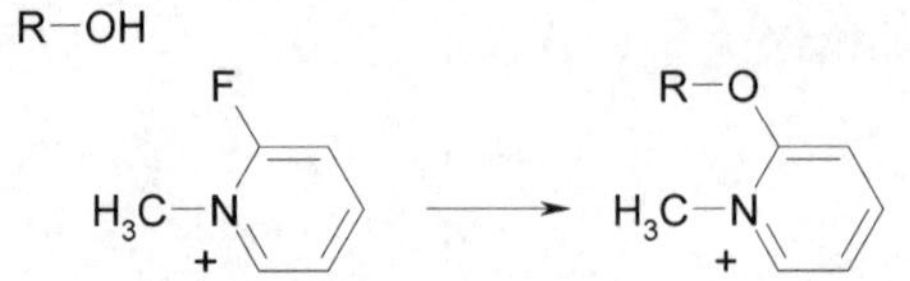

Figure 6.1 Derivatisation of the terminal –OH group on a fatty alcohol or an
ethoxylate derivative with Pyr+.

Table 6.1 Key ions (m/z) used in the identification of Pyr+ derivatised
polyethoxylates of fatty alcohols.

x in EO_x	C_{12}	C_{13}	C_{14}	C_{15}	C_{16}	C_{18}
0	278	292	306	320	334	362
1	322	336	350	364	378	406
2	366	380	394	408	422	450
3	410	424	438	452	466	494
4	454	468	482	496	510	538
5	498	512	526	540	554	582
6	542	556	570	584	598	626
7	586	600	614	628	642	670
8	630	644	658	672	686	714
9	674	688	702	716	730	758
10	718	732	746	760	774	802
11	762	776	790	804	818	846
12	806	820	834	848	862	890
13	850	864	878	892	906	934
14	894	908	922	936	950	978
15	938	952	966	980	994	1022
16	982	996	1010	1024	1038	1066
17	1026	1040	1054	1068	1082	1110
18	1070	1084	1098	1112	1126	1154

After Dunphy et al.[125]

components typically up to EO$_{20}$. The most sensitive detector configuration for
such a suite of compounds is single ion monitoring (SIM). A list of the principal
ions of each species is presented in Table 6.1.

The limits of detection for individual homologues in the study of Dunphy
et al.[125] ranged from $0.1\,ng\,l^{-1}$ to an estimated $22\,ng\,l^{-1}$ but most values were
between 3 and $5\,ng\,l^{-1}$. When these values are summed, it is possible to calculate the limit of quantitation for all alcohol ethoxylates in effluent samples; a
value of $1.7\,\mu g\,l^{-1}$ was suggested for the commercial mixtures of these ethoxylates used in their study. An important limitation in this concept is that the
expected concentrations are unequal such that simple summations of limits of
quantitation should be carefully applied. Morrall et al.[84] further showed that
these limits are sample-dependent as well.

6.3 Environmental Samples

Due to the non-polar nature of fatty alcohols either as free alcohols or as wax esters, the compounds will be present in the sedimentary phase either in suspension or settled.[126] Therefore, most analyses concentrate on this medium when determining environmental concentrations. The association of the compounds within the sediment is not homogenous and several distinct phases may exist. The method chosen for extracting these compounds will, therefore, determine what components are quantified.

The analysis of fatty alcohols in environmental samples falls into two camps: those methods that extract directly into a non-polar organic solvent and those that saponify the sediment directly. The schemes for these two methods can be seen in Figure 6.2. The principal division is in the use of KOH for saponification directly on the sediment or only after lipid extraction with DCM/MeOH. The different routes will yield different values and profiles as the fatty alcohols may be associated with different matrices in the sediment, which will change with source and age.

The "extractable" component is removed by dissolving it into either a DCM or chloroform mix with 1 : 1 methanol (MeOH) after the methods of Folch *et al.*[127] These extractable lipids may then be saponified to break any ester linkages leaving the free lipid in solution, sometimes as its sodium or potassium salt. In the case of waxes, a fatty acid and a fatty alcohol are produced (Figure 6.3). These may then be extracted separately from each other. Typically, the neutral lipids will include the sterols and fatty alcohols and these may be extracted directly into a non-polar solvent such as hexane.[128] The fatty acids may also be extracted, if required, by titrating the KOH/MeOH fraction back to an acid pH with HCl. These polar lipids may then be extracted into 9 : 1 hexane–diethyl

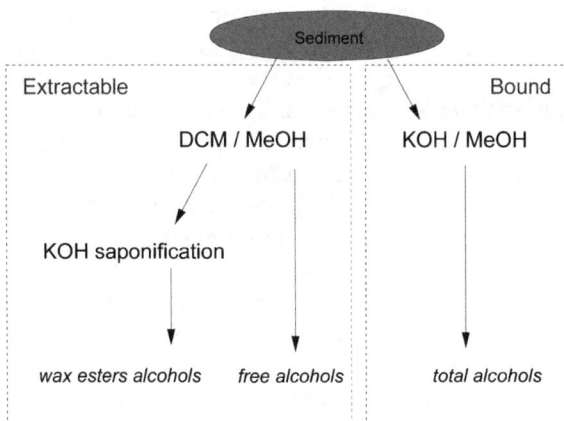

Figure 6.2 Typical extraction protocols for fatty alcohols. The major division is whether the sediment sample is treated directly with KOH to remove all bound compounds or whether it is used only after solvent extraction.

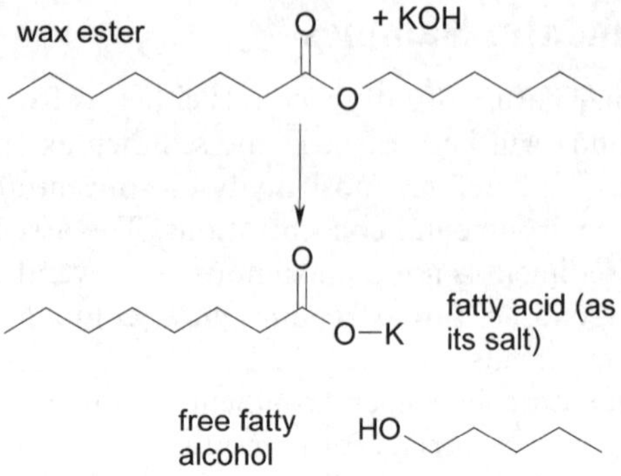

wax ester

Figure 6.3 Saponification process leaving free fatty alcohols in solution.

ether[128] or 1 : 4 DCM–hexane.[113] The alcohols and sterols are then generally derivatised to form the trimethylsilyl (TMS) ethers. Several reagents are available for this purpose, the most commonly used being BSFTA (bis(trimethylsilyl) trifluoroacetamide). The free alcohols may be directly derivatised with BSTFA and analysed in the same manner as other lipids.

The method for bound (wax ester, geolipids, *etc.*) fatty alcohols involves direct saponification of the sediment with alkaline methanol. Moist sediment or biological material is boiled in 6% KOH in methanol (w/v) for approximately 4 hours. Unpublished work by S.M. Mudge has shown that this is sufficient for >95% extraction of the bound sterols. After allowing the extract to cool, the solids should be separated from the liquor by either centrifugation or filtration. If using centrifugation, glass centrifuge tubes of nominally 40 ml capacity should be balanced within 1g and spun at 2500 rpm for ~5 min or until the solids have separated from the liquor. For improved recovery, the solids can be re-suspended in methanol, re-centrifuged and the liquid combined with the initial extract.

The clear liquor from sediment extractions, ranging in colour from pale yellow to dark brown, can be poured into a glass separating funnel using a glass funnel to aid in the transfer. Addition of 20–30 ml of hexane to the liquor and shaking will extract the non-polar lipids that now include any free fatty alcohols originally present plus bound (wax derived) compounds. The polar compounds including the fatty acids will remain in the alkaline methanol. As well as the fatty alcohols, the sterols and PAHs will preferentially partition into the hexane.[56] Derivatisation is by use of BSTFA as above.

The most useful instrumental method for fatty alcohols from environmental samples is gas chromatography–mass spectrometry (GC-MS). The analytical column should be of the DB-5, HP-5, BPX-5 variety although better baselines have been seen using a high temperature column such as the HT-5 of SGE (Scientific Glass Engineering, Australia). The temperature programme needs to

Table 6.2   A list of the key fragments (M^+–CH_3) for fatty alcohol identification by GC-MS analysis. The *iso* and *anteiso* branched components are also included with the parent C_n *m/z* values.

Carbon number	*m/z*	Carbon number	*m/z*
10	215	20	369
11	229	21	383
12	243	22	397
13	257	23	411
14	271	24	425
15	285	25	439
16	299	26	453
17	313	27	467
18	327	28	481
19	341	29	495
20	355	30	509

go to about 360 °C to ensure removal of all compounds from the column and at these elevated temperatures increased column bleed can become troublesome with some columns. Typical column lengths are 30 to 60 m and the best separations can be seen with narrow bores and thin films (0.25 mm and 0.1 μm).

A typical temperature programme should start at 60 °C, increasing at 15 °C min^{-1} to 300° C, then at 5° C min^{-1} to a maximum of 360° C. Other gradients are possible and may be recommended if the sterols and PAHs are to be quantified as well. The mass spectrometer is best configured for electron impact ionisation at 70 eV and a mass scan range of 45–545 *m/z* per second. If better detection limits are required, the mass spectrometer may be operated in the SIM mode using the fragments of M^+–CH_3 as the identifier. A list of key ions is shown Table 6.2.

An example of a GC trace with selected fragments for the homologous series can be seen in Figure 6.4. The major ions from Table 6.2 are shown for the C_{14}–C_{18} compounds above the total ion count (TIC). Examples of mass spectra for alcohols from environmental samples are shown in Figure 6.5.

6.4   Inter-laboratory Comparisons

No reports of inter-laboratory comparisons have been found. It is possible that since these compounds are not routinely reported, such a comparison has not yet been conducted. Considering the different extraction routes possible (nonpolar solvent then saponification *versus* direct saponification into a polar solvent), effort may need to be directed in this direction if comparisons between different locations are to be made. Effort may need to be directed towards such a comparison exercise.

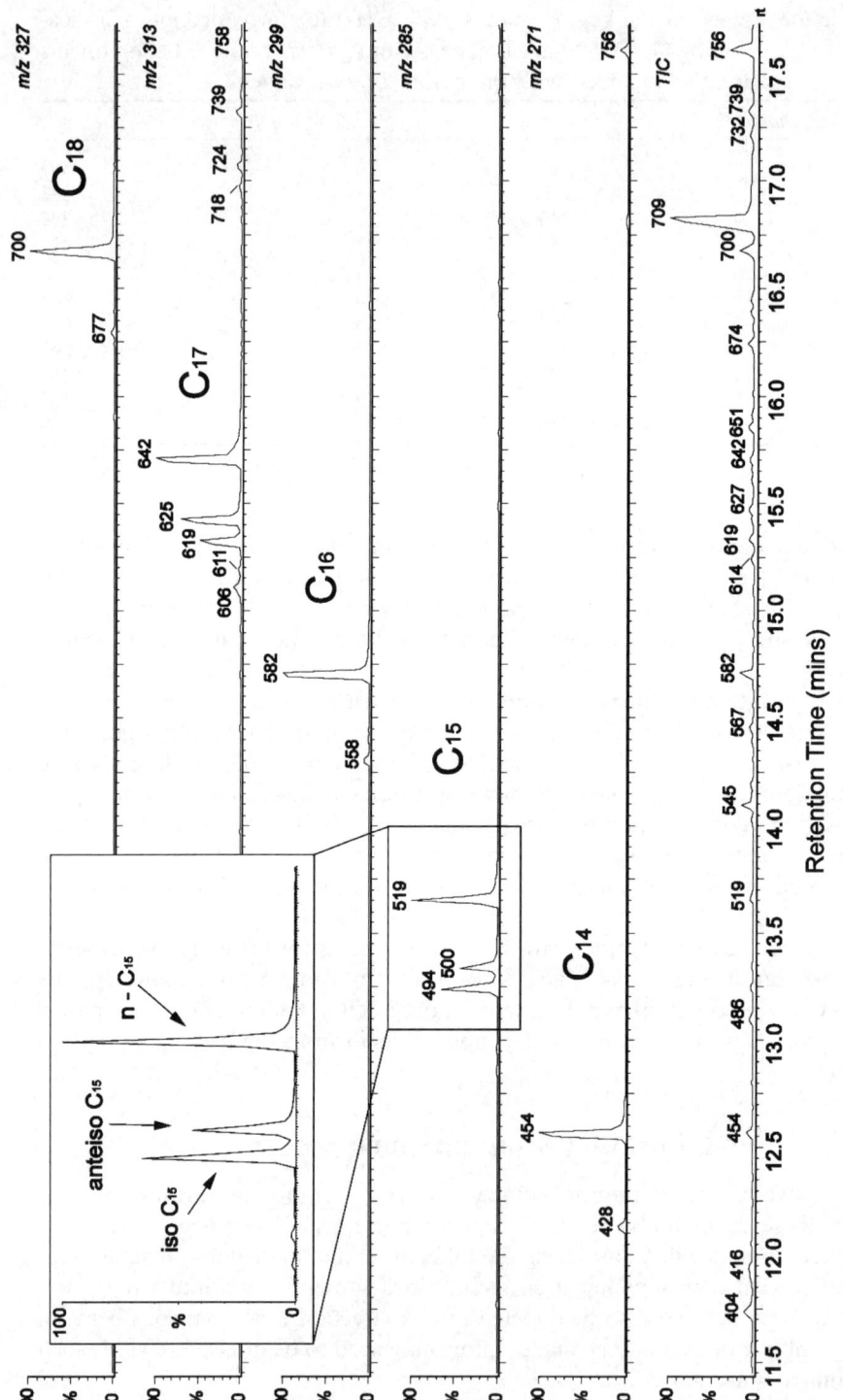

Figure 6.4 GC traces for an alkaline saponification of surface sediment (Menai Strait, North Wales) with key ions for a range of fatty alcohols. The expanded view shows the branched chain compounds for the odd carbon numbered species.

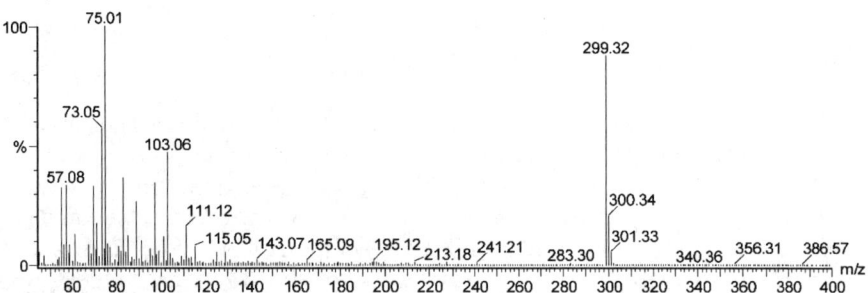

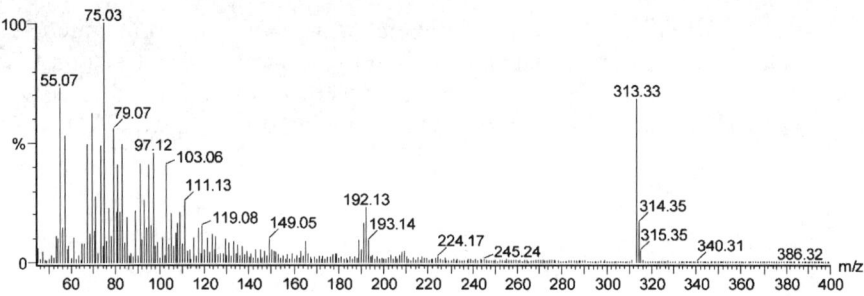

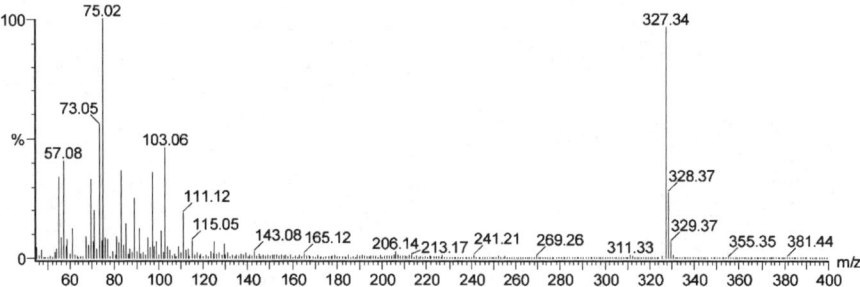

Figure 6.5 Mass spectra of the C_{16} (299 ion), C_{17} (313 ion) and C_{18} (327 ion) *n*-alcohols. The characteristic ions ($M^+ - 15$) can clearly be seen in each case.

Summary

- Free fatty alcohols may be extracted directly with a mid-polarity solvent (*e.g.* chloroform–methanol), derivatised with BSTFA and analysed by GC-MS.
- Wax esters and other bound fatty alcohols will need to be saponified before analysis. This may take place either directly on the solids or after solvent extraction.
- Polyethoxylates need a different extraction method using Pyr+ and analysed by LC-MS.
- The different associations of the fatty alcohols requiring different extraction methods means that when environmental samples are analysed, more than one fraction may need to be analysed if the detergent-derived polyethoxylate alcohols are to be placed in context.

CHAPTER 7

Environmental Concentrations

What's out there? Has it changed much (core data) and are there "patterns" in the distributions that tell us about sources or degradation pathways?

This chapter sets out to describe the environmental concentrations of fatty alcohols in a range of environments around the world. It is approached by initially reviewing a range of studies that have measured such compounds and then synthesising these data to determine if patterns exist that may be explained by geographic location or proximity to anthropogenic sources. One medium that falls between the concentrations in organisms (Chapter 3) and the concentrations in inert environmental samples (this chapter) is faecal matter, either from land drainage or through sewage treatment plants (STPs). These sources have been considered in Chapter 5 on environmental transformations.

Compared to other groups of compounds, fatty alcohols have received relatively little attention in environmental studies. This may be due to the occurrence of better lipid biomarkers for determining a source (*e.g.* the sterols[56]) or environmental conditions[129,130] or more sensitive indicators of degradation (*e.g.* the fatty acids[108]). On the positive side, however, fatty alcohols are extracted along with sterols and other biomarkers and so many studies do generate these data. Of these, few use the alcohols in a diagnostic manner and the data are only touched upon in many papers or remain in unpublished masters or doctoral theses. As many original datasets as possible have been reviewed. For each study, a brief overview of the purpose is given followed by some concentration data and distributions. Relatively very few measurements have been found for terrestrial environments compared to marine and freshwater systems.

The study locations are shown in Figure 7.1, with details given in Table 7.1. The UK locations are shown in Figure 7.2, with details given in Table 7.2.

The majority of studies reporting the concentrations of environmental fatty alcohols investigated sediments or soils. This is directly related to the non-polar nature of these compounds either as free fatty alcohols or as bound wax esters.

Fatty Alcohols: Anthropogenic and Natural Occurrence in the Environment
By Stephen M Mudge, Scott E Belanger, and Allen M Nielsen
© Copyright 2008 ERASM (the joint surfactant environmental research platform of AISE and CESIO) and SDA
Published by the Royal Society of Chemistry, www.rsc.org

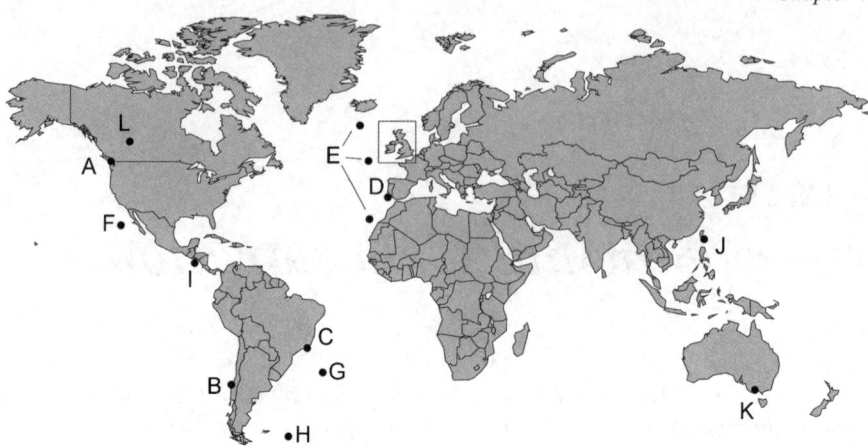

Figure 7.1 Map of study locations with reported fatty alcohol data. Data locations for the UK are shown in Figure 7.2.

In both cases, very small proportions of the totals will be in the dissolved phase; if water samples are taken, the fatty alcohols are present predominantly on the suspended particulate matter.[126]

This is in direct contrast to the occurrence of fatty alcohol polyethoxylates; the long repeating chain of $-CH_2CH_2O-$ dramatically increases the water solubility and will alter the transport of the compounds accordingly. However, there will be ether cleavage, shortening of the ethoxylate chain and production from alcohol sulfates through time in sewage treatment plants.[121] With these processes the water solubility decreases and the compounds will eventually become free fatty alcohols (and consequently less water soluble) and partition on to the solid phase accordingly. Therefore, studies of sedimentary material will contain fatty alcohols derived from both natural and anthropogenic sources but the transport paths may be different.

7.1 The Marine Environment

7.1.1 Victoria Harbour, BC, Canada: Estuarine Surface Sediments (A)

A study was undertaken in Victoria Harbour, BC, Canada (Figure 7.3) to assess the extent of sewage contamination;[131] bacterial studies had suggested the upper reaches were unaffected by treated sewage discharged near the harbour. The study investigated sterols principally but also reported fatty acids and fatty alcohols. GC-MS analysis of the samples identified 37 fatty alcohols in surface sediment samples after alkaline saponification. In these samples, the C_{22} fatty alcohol had the highest mean concentration ($193 \, \mu g \, g^{-1}$) and may have been directly related to the presence of wood chips in the samples from

Table 7.1 Study locations, sample dates and references for fatty alcohol data relating to the map in Figure 7.1.

Code	Location	Country	Sample type	Sampling period	Reference
A	Victoria Harbour, BC	Canada	Estuarine surface sediments	July 1998	131
B1	Concepción Bay	Chile	Marine surface sediments	January 2000	132
B2	San Vicente Bay	Chile	Marine surface sediments	January 1998	133
C	Rio de Janeiro	Brazil	Embayment surface sediments	December 2000	Mudge and Neto (unpublished)
D1	Ria Formosa lagoon	Portugal	Lagoonal surface sediments	June 1995	Mudge (unpublished)
D2	Ria Formosa lagoon	Portugal	Lagoonal sediments	June 2002	126
D3	Ria Formosa lagoon	Portugal	Shallow lagoon core	June 2001	111
E	Eastern North Atlantic		Oceanic cores	Summers of 1990–1991	134
F	San Miguel Gap, California	USA	Oceanic cores	Early 1980s(?)	135
G	Rio Grande Rise (516F of leg 72 ODP[a])	Brazil	Oceanic cores	Early 1980s(?)	136
H	Falkland Plateau (511 of leg 71 ODP[a])	South Atlantic	Oceanic cores	Early 1980s(?)	136
I	Guatemalan Basin (legs 66 and 67 ODP[a])	Central America	Oceanic cores	Early 1980s(?)	136
J1	Continental slope, southwest of Taiwan	Taiwan	Oceanic surface sediments	1996(?)	117
J2	East China Sea, north of Taiwan	Taiwan	Oceanic surface sediments	May–June 1999	55
K	Pasture land, southern Australia	Australia	Freshwater runoff	November 2000	137
L	Prairie zone soils, Alberta	Canada	Terrestrial soils	2003(?)	138

[a]ODP, Ocean Drilling Programme.

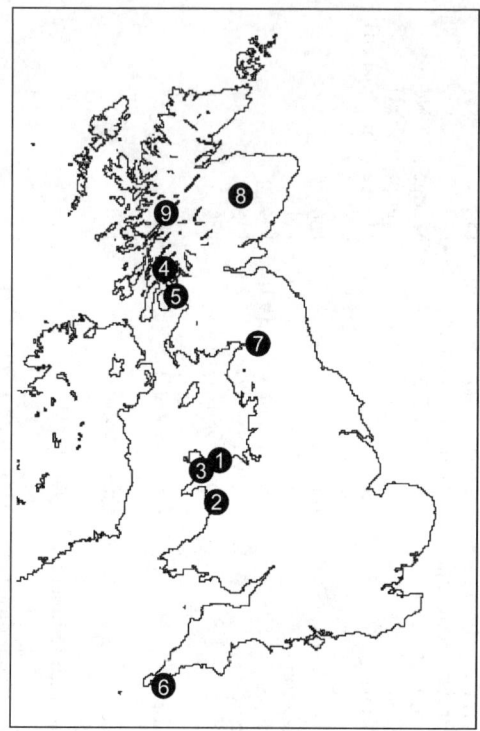

Figure 7.2 Locations of fatty alcohol data collected in the UK.

Table 7.2 UK data locations.

Code	Location	Country	Media	Sampling period	Reference
1	Conwy estuary	Wales	Estuarine core	April 1997	110
2	Mawddach estuary	Wales	Estuarine surface sediments	June 1996	110
3	Menai Strait	Wales	Marine surface sediments	1996–2005	Mudge (unpublished)
4	Loch Riddon	Scotland	Marine core	May 1998	110
5	Clyde Sea	Scotland	Marine surface sediments	May 1998	110
6	Looe Pool	England	Brackish lagoon		139
7	Bolton Fell	England	Peat core		69
8	Lochnagar	Scotland	Lake core		140
9a	Loch Lochy	Scotland	Freshwater core	November 2000	109
9b	Loch Eil	Scotland	Marine core	November 2000	109

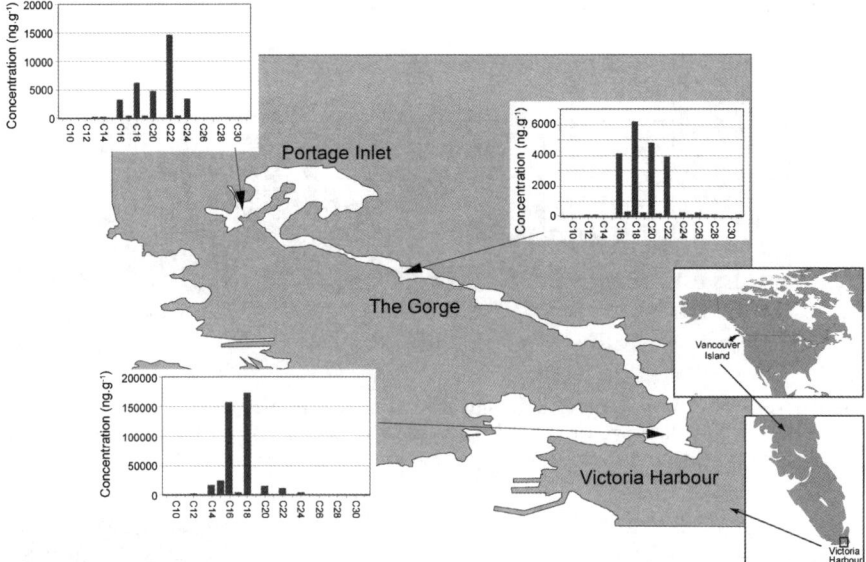

Figure 7.3 Sampling location in Victoria Harbour cited in Mudge and Lintern[131] and the distribution of fatty alcohols in selected surface sediment samples. These show the change from short chain moieties near the seaward end to longer chains in the upper reaches.

adjacent wood processing industries. The distribution between long and short chain fatty alcohols changes between sites in the system, with more short chain fatty alcohols near the marine end (C_{14}), longer chain fatty alcohols in upper reaches (C_{22} and C_{24}) and a mixture of mid-length compounds in the intervening sections (C_{16}, C_{18} and C_{20} dominant) (Figure 7.3). The results are not as clear-cut as implied since terrestrial organic matter in the form of by-products from wood processing was found at all sites.

This unusual heterogeneity can be seen in the traditional marker ratios such as C_{24}/C_{16} where the longer chain alcohol is derived from terrestrial plants and the latter one from marine organisms. Figure 7.4 shows this ratio for samples organised according to their distance from the sea; samples to the left are closest to the marine end while those at the right are from the upper reaches. There is no gradient in this marker which shows a wide range of values with no obvious trend, probably due to the wood chip distribution. Concentrations of these fatty alcohols were significantly greater than in many of the other similar sediment studies where wood chips were absent.

7.1.2 Concepción Bay, Chile (B1); San Vicente Bay, Chile (B2)

In a study of two anthropogenically contaminated bays in Chile (Figure 7.5), subtidal surface sediments were collected and extracted by alkaline saponification.[141]

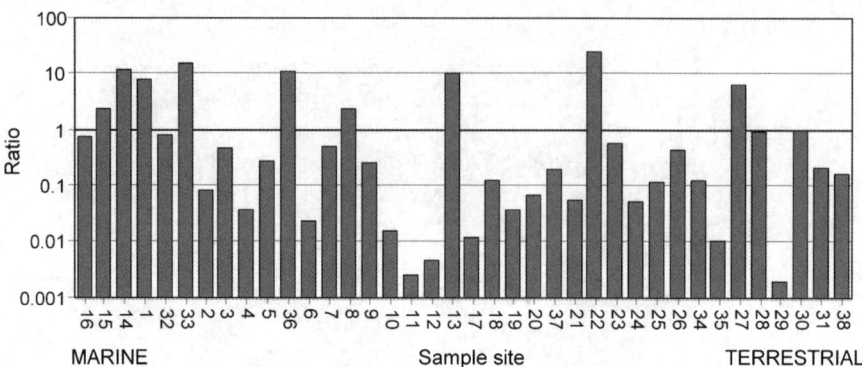

Figure 7.4   Ratio of C_{24} to C_{16} in surface sediment samples from Victoria Harbour. The data are organised by increasing distance from the sea (from left to right) and the ratios are presented on a $\log_{10}$ scale.

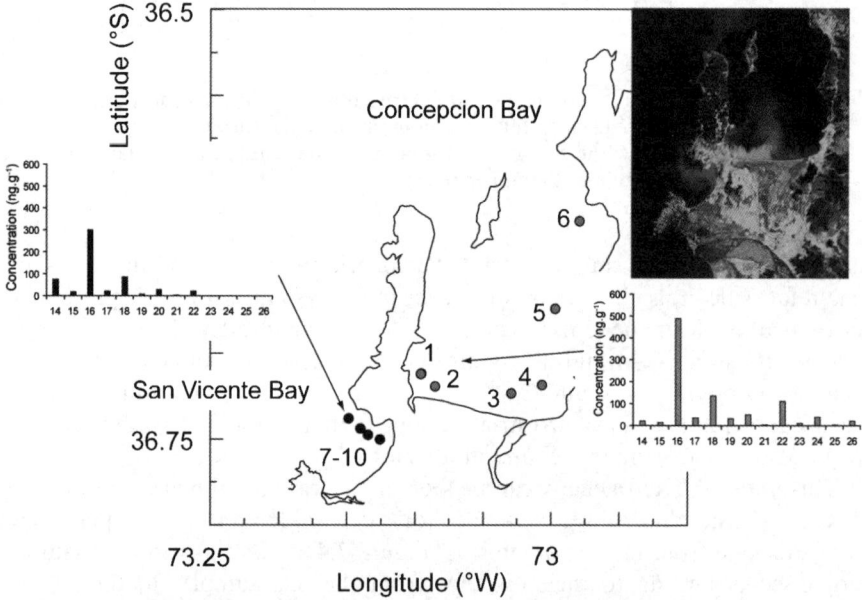

Figure 7.5   Sampling locations for sub-tidal surface sediments in Chile[141] with typical fatty alcohol chain length profiles for each bay shown. The inset is a LandSat image taken in mid-summer and clearly shows waste discharges.

Fatty alcohol data were collected together with data for alkanes, fatty acids and sterols. Sediment fatty alcohol concentrations were similar to those found elsewhere, although the profile was not quite as expected (Figure 7.5).

All of the samples were relatively depleted in long chain ($>C_{20}$) comounds. These waters are fully saline but Concepción Bay does receive terrestrial runoff. The surrounding areas are densely wooded and long chain compounds might be expected. However, as Figure 7.5 indicates, the short chain C_{16} dominates in both bays indicating a strongly marine faunal signature. The lack of a terrestrial signature both in the fatty alcohol and sterol data prompted a further investigation of this bay.[132] Sub-tidal surface sediments were again collected but only from Concepción Bay. These Concepción Bay sediments were soft, black, reducing muds, rich in organic matter with a thin overlying flocculent layer with anoxic conditions prevailing at the sediment–water interface during most of the year.[142] Domestic wastewater from several small outfalls, the Andalien River and from two pipes, has long been discharged in to Concepción Bay, although a primary sewage treatment plant has been built to serve the nearby coastal city of Penco-Lirquen.[143] The effluent discharge pipe extends 1300 m into the bay; the depth of the water column at the disposal site is 25 m, with a discharge rate of 95 litres per second, 14 m above the seabed. It is located near site 4 in Figure 7.5.

All samples had high concentrations of phytol. This compound was the most abundant fatty alcohol with a range of $0.1–98.6\,\mu g\,g^{-1}$ dry weight (DW). This suggested that the study area had high productivity of marine microalgae. Straight chain fatty alcohols between C_{12} and C_{30} were identified in the samples together with several short chain, odd carbon number branched compounds. In general, sites close to the sewage disposal had the highest total concentrations, especially in the top 0 to 15 cm of the sediment, ranging from 0.1 to $10\,\mu g\,g^{-1}$ DW (excluding phytol). Small amounts of the branched fatty alcohols *iso*C15:0, *iso*C17:0 and the corresponding *anteiso* forms were also found; these fatty alcohols had a clear prevalence at some sites in the top 0 to 15 cm of sediment (9.1 to 38.4% of total *n*-alcohols).

In all samples there was an even-over-odd dominance in the *n*-alcohols, with fatty alcohols C_{16}, C_{26} and C_{28} abundant in all samples. The ratio between odd chain length and even chain length of saturated fatty alcohols $(\Sigma C_{13}–C_{29})/(\Sigma C_{14}–C_{30})$ has been used as an indicator of bacterial activity; in Concepción Bay this ratio was essentially small at most sites, although the range was from 0.01 to 0.52. Although most ratios had low values, sites furthest from the discharge point had the highest values of this bacterial indicator. Bacteria have signatures with odd chain and branched fatty alcohols, although the profile changes depending on the species present.

The ratio between long and short chain saturated fatty alcohols can be used to indicate the terrestrial input into the system; values greater than 1.0 are indicative of enhanced long chain fatty alcohol concentrations potentially from terrestrial sources. However, short chain compounds are more readily degraded than long chain ones and this may alter the ratio. The distribution of this ratio $(\Sigma C_{19}–C_{30})/(\Sigma C_{12}–C_{18})$ in Concepción Bay showed that sites close to the sewer (2 to 6 in Figure 7.6) had the highest values indicating mainly terrestrial sources; this was also true of the sterol biomarker ratio, although this value did not exceed 0.4. The compound β-sitosterol (24-ethyl cholesterol) is produced by terrestrial plants while many marine animals synthesise cholesterol instead. This may indicate a possible terrestrial runoff source associated with the Andalien River or associated with the sewage input itself.

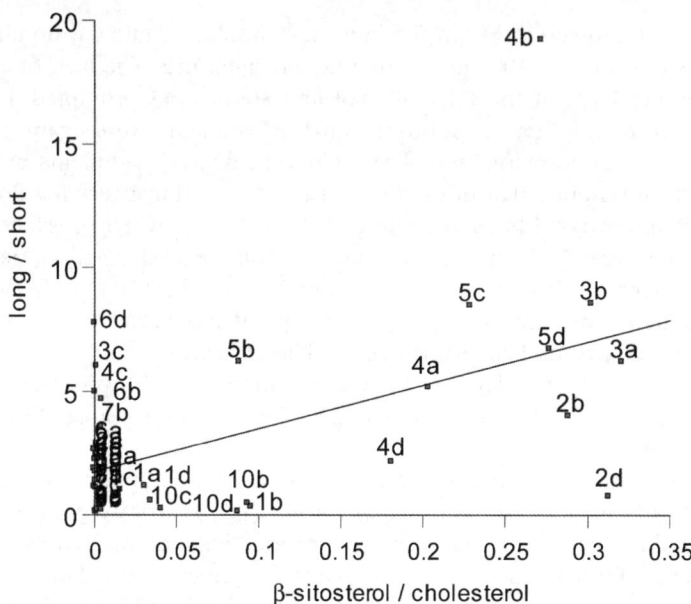

Figure 7.6   Long ($>C_{18}$)/short ($<C_{19}$) ratio plotted against the sterol marker for terrestrial organic matter (β-sitosterol/cholesterol).

These data do not suggest that the fatty alcohols are a good biomarker but the same could be said of the sterols. However, the sterols are relatively stable in the marine environment and have been used as indicators of organic matter sources for several years. In this case, the strongly reducing nature of both the waters and sediments may lead to altered chemical behaviour compared to other places. The satellite image of the region in Figure 7.5 shows the plumes of organic input that have considerable effect on the water quality.

7.1.3   Rio de Janeiro: Surface Sediments in a Contaminated Bay (C)

The city of Rio de Janeiro is located on the South Atlantic coast of Brazil and around Guanabara Bay. The bay has a surface area of $384 \, km^2$ and a perimeter of 131 km. The mean water depth is 5.7 m and has a maximum depth of 30 m in the entrance channel such that 50% flushing of water in the bay takes 11.4 days. To the north, it is bordered by $90 \, km^2$ of fringing mangroves, of which $43 \, km^2$ is an "Environmentally Protected Area". The bay receives wastes and runoff from the catchment; sewage from nearly 8 million people discharges directly into the bay with little treatment and there is considerable waste from industry and shipping. The port area is continuously dredged.

In an unpublished study, Mudge and Neto collected surface sediments with a grab from a small boat. Samples were analysed by alkaline saponification. The ratio C_{22}/C_{16} is presented spatially in Figure 7.7 as a classed posting together with the C_{16} concentration as point labels.

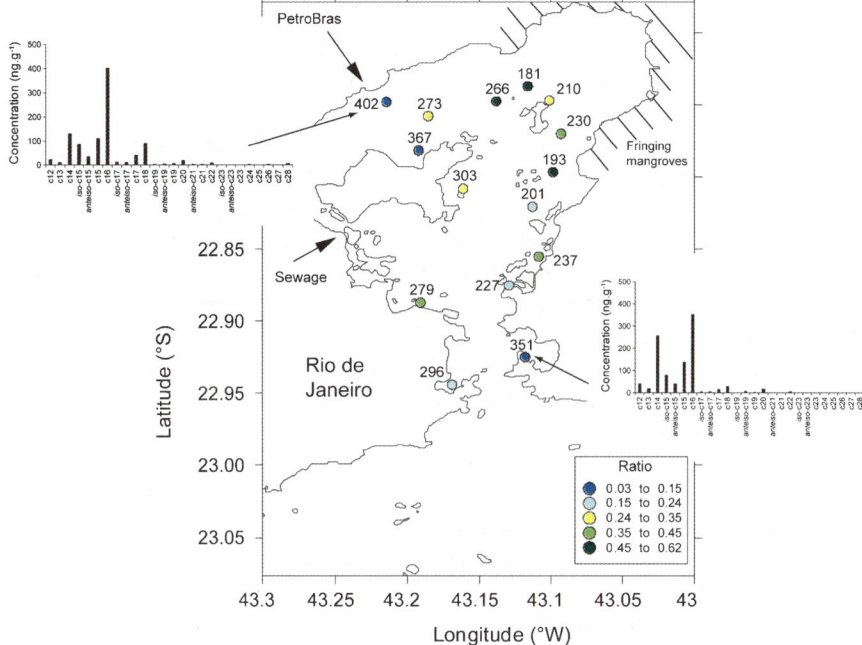

Figure 7.7 Spatial distribution of the C_{22}/C_{16} ratio in Guanabara Bay, Brazil, shown as a classed posting. Each sampling site is also labelled with the C_{16} concentration in $ng\,g^{-1}$. Examples of the fatty alcohol profile at two locations are also included.

The highest values of the ratio indicating terrestrial organic matter are located in the northeast portion of the bay adjacent to the mangrove swamps. This would reflect the input of terrestrial matter that may be degrading in the bay. The lowest values are toward the outer reaches most influenced by the sea and in the northwest corner near the STP outfalls from the city and the large oil refinery belonging to PetroBras. The reason for these sites being high in C_{16} is not immediately obvious except that it also has the highest sterol sewage indicator (5β-coprostanol). A comparison of the fatty alcohol profiles at two of the sites shows how similar they are despite the northern site having the greatest 5β-coprostanol/cholesterol ratio, an indicator of human sewage. The southern site had a sewage input almost seven times less by this ratio. It is possible that a component of these alcohols may have been derived from an anthropogenic or sewage source rather than a natural plant one.

7.1.4   Ria Formosa Lagoon: Surface Sediments (D1)

The Ria Formosa lagoon in southern Portugal is one of the largest lagoons in Europe with a length of ~55 km. The system receives inputs from several sources including domestic sewage from the towns and cities of the coastal zone, especially Faro; there is terrestrial runoff of nutrients as the hinterland is intensively farmed but riverine input is small and confined to a few rivers of which only one runs throughout the entire year. There have been several studies of the system for a range of contaminants including PCBs,[144] sterols,[112] fatty acids,[113] nutrients,[145] oxygen[146] and metals.[147] As part of an earlier study, Mudge and co-workers collected surface sediments and analysed these after alkaline saponification for sterols and fatty acids, the results of which have been reported in the literature. However, they also measured aliphatic and aromatic hydrocarbons and fatty alcohols. These data have yet to be reported.

Fatty alcohols ranging from C_9 to C_{32} were measured together with a number of branched chain compounds: C_{16} was present in the highest concentration, with a maximum value of $1348\,\mathrm{ng\,g^{-1}}$ DW. An example of the profile for two different sites can be seen in Figure 7.8. Site 35 is close to the major sewage outfall for the city of Faro and has high faecal sterol markers,[112] while site 43 is close to the wide sandy entrance to the North Atlantic at Armona (Figure 7.9). The data from Figure 7.8 show how there are more long chain ($>C_{20}$) compounds in the inner site influenced by the sewage discharge than the outer site, which is marginally enriched in the short chain compounds compared to site 35.

It is possible to view these data spatially to determine the regions where compounds are associating. Figure 7.9a presents the sum of the C_9–C_{14} fatty alcohol concentrations, generally assumed to be the marine component, while Figure 7.9b shows the ratio of C_{24}–C_{32}/C_9–C_{14}, high values of which should indicate the most likely terrestrially derived organic matter accumulation sites. Highest concentrations (blue circles) of the short chain compounds can be seen in the Armona inlet which supports a community of sand dwelling organisms including microalgae;[113] low values (red circles) can be found in a diversity of habitats: in the river (site 14), in the main navigation channel which is routinely

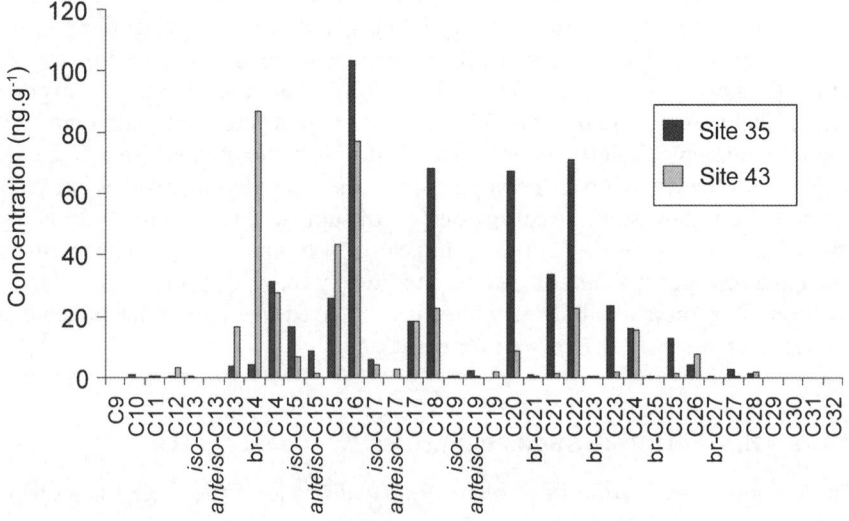

Figure 7.8 Fatty alcohol profile for two contrasting sites in the Ria Formosa lagoon.

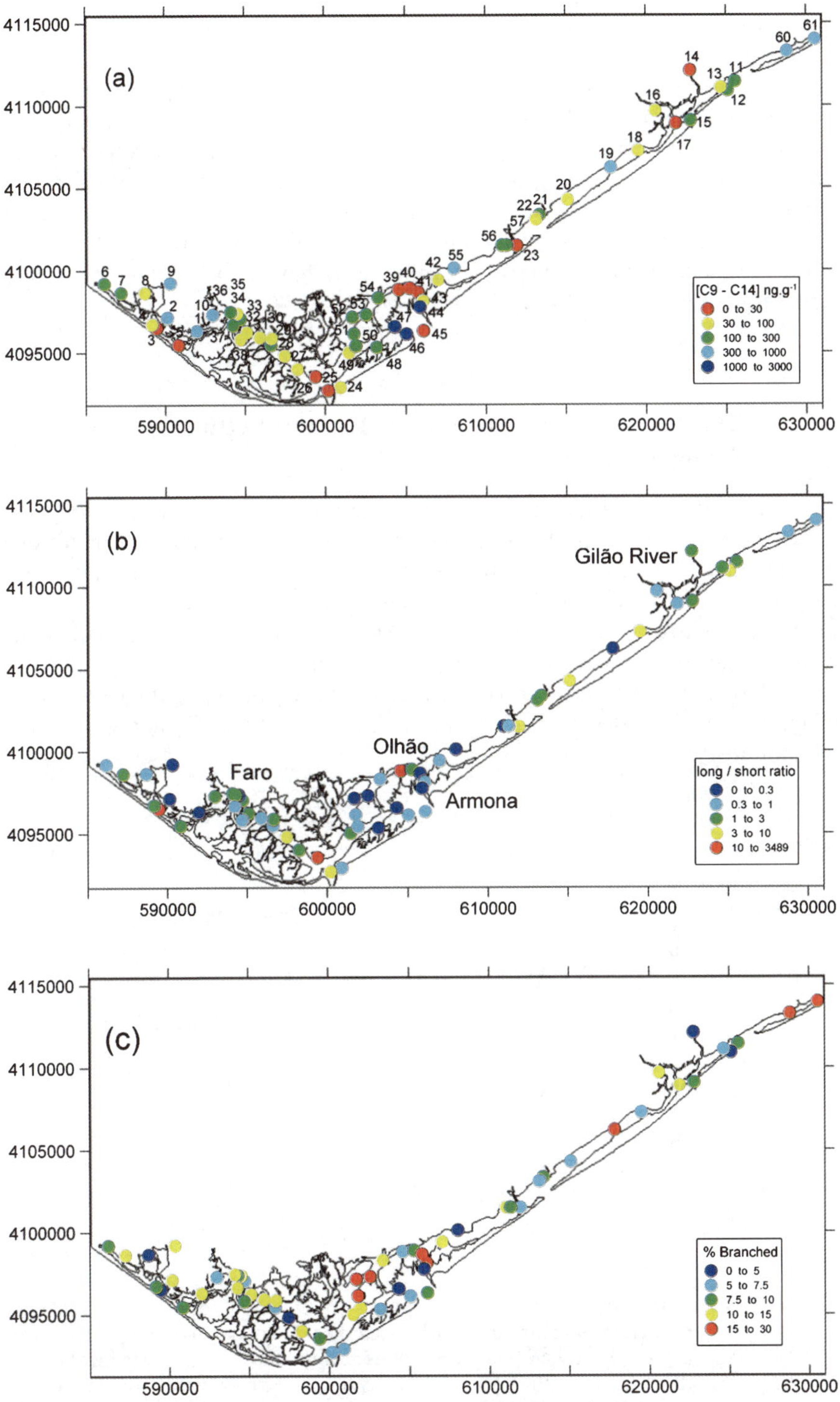

(a)

14
13 11
60 61
16
12
15
19 18
17
20
22 21
56 57
55 23
42
6 7 8 9
1 36 35 39 40
2 34 33 41
10 32 53 55 43
3 37 52 50 44
38 51 47 45
49 28 48 46
27 26 25 24

[C9 - C14] ng.g⁻¹

- 0 to 30
- 30 to 100
- 100 to 300
- 300 to 1000
- 1000 to 3000

(b)

Gilão River

Olhão
Faro
Armona

long / short ratio

- 0 to 0.3
- 0.3 to 1
- 1 to 3
- 3 to 10
- 10 to 3489

(c)

% Branched

- 0 to 5
- 5 to 7.5
- 7.5 to 10
- 10 to 15
- 15 to 30

Figure 7.9 (a) Concentration of short chain fatty alcohols, (b) long/short chain ratio and (c) percentage branched chains across the Ria Formosa, Portugal.

dredged (sites 24–26) and near seagrass beds (sites 3–4 to the west and 39–40 in the centre of the lagoon).

One of the measures of bacterial activity in any system is the percentage of all fatty alcohols in a branched configuration. In these samples, this measure can be seen in Figure 7.9c. The highest values were associated with sites that are known to be areas of fine-grain sediment accumulation and the sediments are anaerobic. As with other studies, the fatty alcohols in this form do not appear to be very good indicators of sewage inputs unlike the sterols.[112] However, when examined using multivariate techniques (Chapter 9), greater discrimination between sites can be seen.

7.1.5 Ria Formosa Lagoon: Suspended and Settled Sediments (D2)

A further study was undertaken to determine the sources and transport paths of sewage-derived materials in this lagoon. The data are reported in Mudge and Duce.[126] Key factors in this study were the collection and analysis of suspended particulate matter and its relationship with settled sediments and potential origins of the organic matter. The concentrations of fatty alcohols were significantly greater in the potential source materials (*e.g.* sewage disposal sites) than either the suspended particulate matter or the settled sediments (Figure 7.10). This decrease is not surprising given the labile nature of these compounds.

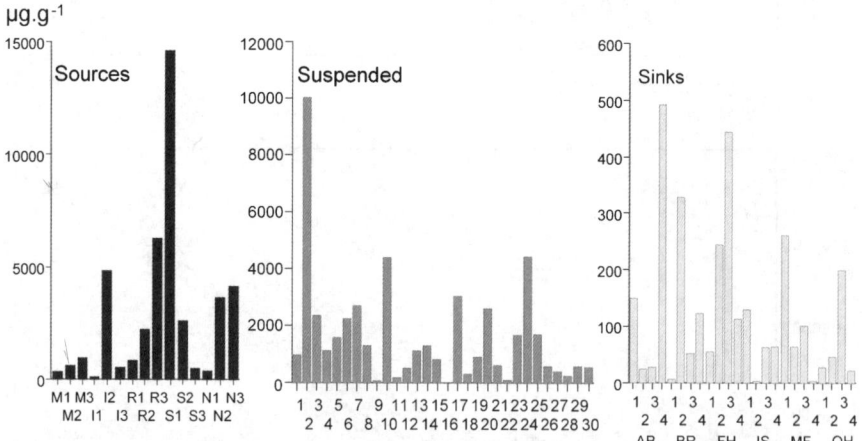

Figure 7.10   Total fatty alcohol concentrations in potential source materials, suspended sediments and settled sediments (sinks) in the Ria Formosa lagoon, Portugal. (Data from Mudge and Duce[126] where the locations in the lagoon to the west of Faro can be seen). Numbers on the *x*-axes correspond to sampling site numbers.

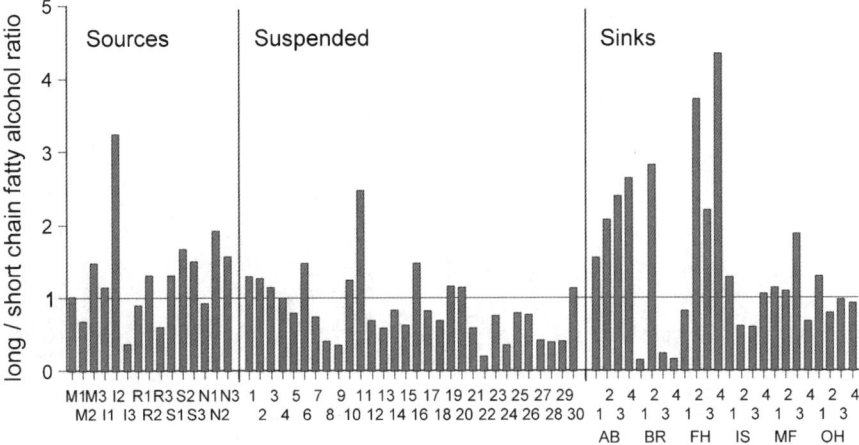

Figure 7.11   Ratio of long chain and short chain fatty alcohols in the sediments from the Ria Formosa. Numbers on the *x*-axes correspond to sampling site numbers.[126]

The fatty alcohols may also give source information based on the ratios of key components such as long chain (terrestrial) compounds and short chain (marine) compounds. Plots of these measures suggested a subtle change in profile between the source and suspended matter and the settled sediments where the relative proportion of the long chain compounds increased (Figure 7.11). This is due in part to the relative resistance of these compounds to degradation.

The odd chain compounds may also be used to indicate bacterial sources. This ratio is not particularly diagnostic in this situation, as Figure 7.12 shows; all source types and all other sediments have mixed ratios with some showing high values of the ratio. In these cases, the bacterial biomass may be high, although there is no clear trend across the locations.

7.1.6   Ria Formosa Lagoon: Shallow Core from Intertidal Sediments (D3)

In a study of the potential environmental changes that are recorded in the sediments, Unsworth[111] collected a core from the productive Ria Formosa lagoon and extracted sterols, fatty alcohols and PAHs. The concentration data for key compounds are shown in Figure 7.13. The region is regularly bioturbated in the search for shellfish and few undisturbed core sites were available. The short chain compounds are present in relatively low concentrations compared to other sites in the lagoon[126] and decrease with depth as might be expected from their degradability. The longer chain compounds are present in much lower concentrations as there is very little terrestrial runoff into this coastal lagoon. Even so, the concentrations do decrease with depth indicating a degree of degradation with time.

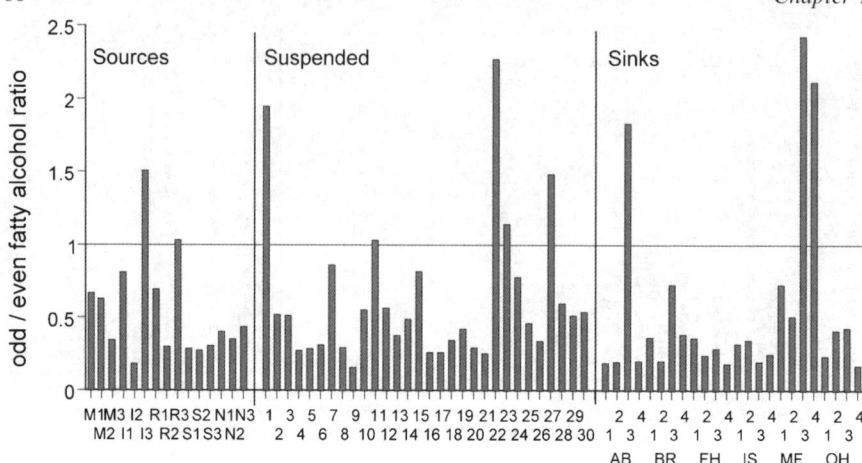

Figure 7.12   Odd chain to even chain ratio for fatty alcohols in the sedimentary matter
from the Ria Formosa. Numbers on the *x*-axes correspond to sampling
site numbers.[126]

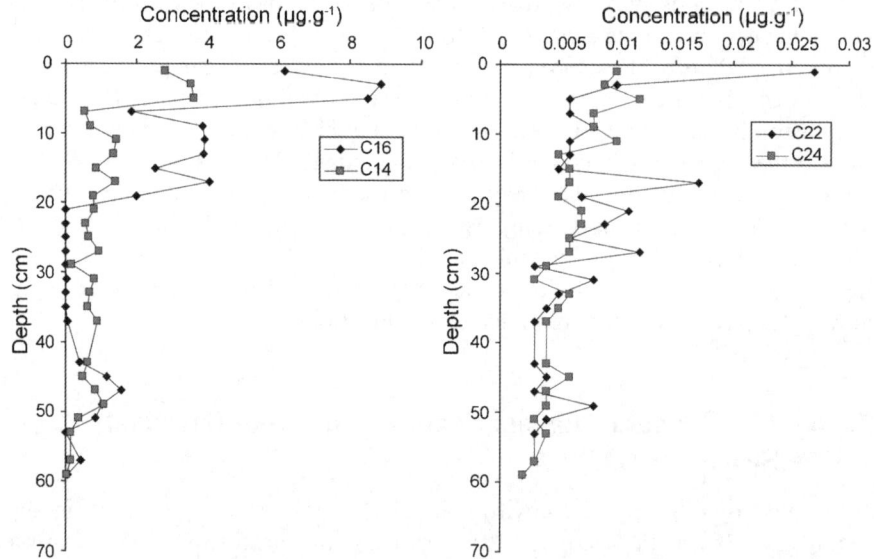

Figure 7.13   Concentration of (left) short chain and (right) long chain fatty alcohols in
a short core (60 cm) in the productive Ria Formosa lagoon, Portugal.
Data after Unsworth.[111]

7.1.7   Eastern North Atlantic (E)

Madureira[134] studied the lipids present within five cores from the eastern North
Atlantic (see Figure 7.1 for approximate locations of Madureira's study). The

results include a suite of fatty alcohols from C_{16} to C_{28} obtained from the sediment cores after alkaline saponification. Examples of the results are presented in Figure 7.14. The distribution of the alcohols at each of the sites is biased toward the long chain moieties with maximal concentrations in the C_{22} to C_{26} length range although the long/short chain ratio does indicate enrichment in the short chain marine compounds near the surface.

If the $\log_e$ values of the C_{16} and C_{26} compounds for an example core are taken, a plot against the accumulation rate of the sediment can give an indication of the degradation rate (Figure 7.15). In this case, the rate of loss of C_{16} is greater than C_{26} as might be expected from the chain length information. Here, the sedimentation rate has been assumed to be 3 mm y^{-1}. Since this rate is only used to compare the relative rates, the exact value is not important.

7.1.8 San Miguel Gap, California: Long Marine Core (F)

McEvoy[135] conducted a study on lipids from a long core collected off the Californian coast. McEvoy first extracted lipids into a DCM–methanol solvent (2 : 1) and then saponified the lipids collected. Therefore, any wax bound fatty alcohols that were extractable in the DCM–methanol will be quantified in this method. Only bound, unextractable with DCM–methanol compounds will not be included.

The results are shown in Figure 7.16. Concentrations near the surface were almost 1000 ng g^{-1} but decreased with depth. By 500 m, no fatty alcohols were detectable, although fatty acids could be recovered all the way down to 1000 m below the surface.

As well as the concentrations decreasing down core, the profile of fatty alcohols also changed reflecting the preferential degradation of short chain compounds. The chain length profile near the surface (Figure 7.16) shows C_{24} to be present in the highest proportion of the total alcohols which shifts towards C_{28} in the 335 m sample. The intermediate sample has a bimodal distribution with some short chain compounds present as well. The short chain compounds may indicate a degree of deep sediment production or a change in the sediment deposition regime when these materials were at the surface. The presence of fatty alcohols down to 335 m in the core indicates that these compounds are well preserved over this time frame. As it is the long chain compounds not the short chain ones that survive, it also indicates the differential degradation rates for the range of environmentally important compounds.

7.1.9 Rio Grande Rise (516F of Leg 72 ODP), Brazil (G)

As part of a series of analyses, Howell[136] analysed the lipids in a core taken from the Rio Grande Rise off the coast of Brazil. This was part of leg 72 of the Ocean Drilling Programme (ODP) cruise to this region. Sediments were extracted by DCM–methanol–hexane (2 : 1 : 1) initially and then saponified with KOH in methanol. The results indicated a very substantial marine signature with a peak

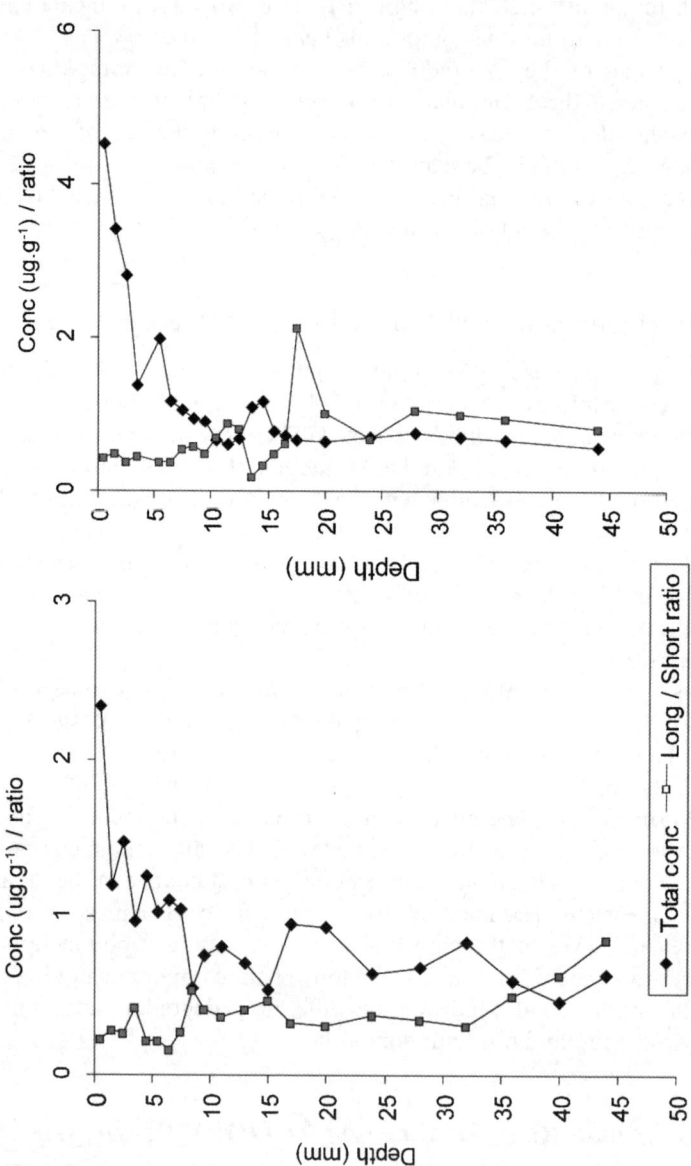

Figure 7.14 Total fatty alcohol concentration and long (C_{22}–C_{28})/short (C_{16}–C_{20}) ratio in two eastern North Atlantic cores: 48° N (left) and 32° N (right). Data after Madureira.[134]

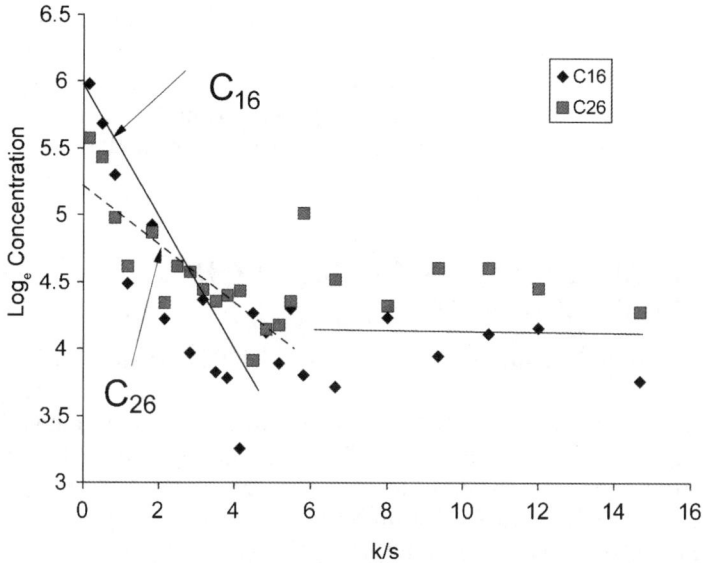

Figure 7.15 The relative degradation rates shown by a plot of the $\log_e$ of the C_{16} and C_{26} fatty alcohols *versus* a sediment accumulation rate (assumed to be $3\,\text{mm}\,\text{y}^{-1}$). Data for site $32°\,\text{N}$ from Madureira.[134]

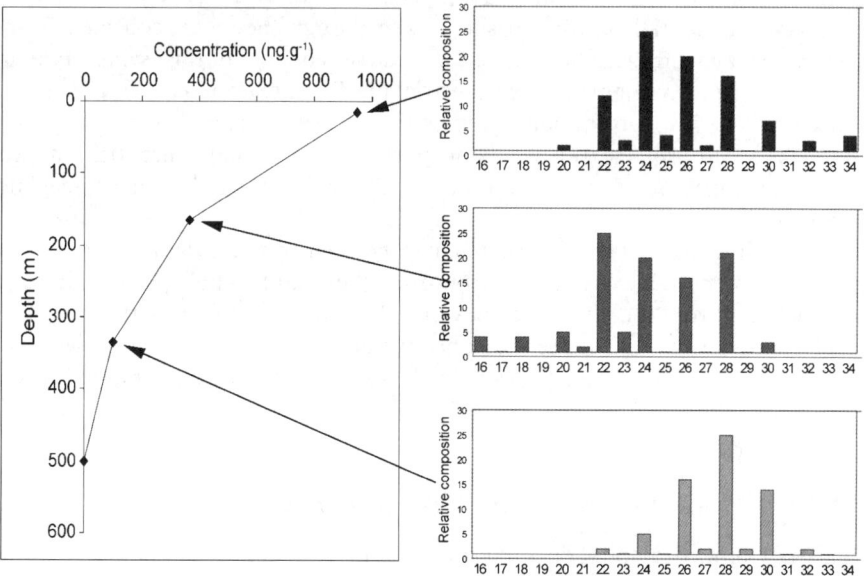

Figure 7.16 Total fatty alcohol concentration in core sediments from the San Miguel Gap, California, with chain length profiles. Data after McEvoy.[135]

occurrence at C_{16} tailing to near zero by C_{22}. This indicates the almost complete absence of terrestrial organic matter with long chain fatty alcohols at this site. The concentration of the C_{16} alcohol is also substantially less ($67\,ng\,g^{-1}$) than observed in more coastal environments (*e.g.* Ria Formosa up to $1348\,ng\,g^{-1}$), indicating a limited *in situ* production or regeneration in the water column before reaching the sediments.

7.1.10   Falkland Plateau (511 of Leg 71 ODP), South Atlantic (H)

Howell also studied a site to the east of the Falkland Islands on the side of Maurice Ewing Bank.[136] After similar extractions and analyses, the results generated a profile even more biased toward short chain compounds than the sample off Brazil (see Section 7.1.9). All four profiles showed a dominance of the C_{12} with low concentrations of long chain alcohols present. This again must represent an entirely marine signature with little or no transport of terrestrial organic matter obvious in the samples.

7.1.11   Guatemalan Basin (Legs 66 and 67 ODP), Central America (I)

A third group of samples from a tropical, near-shore location in the Mid-America Trench off Mexico and Guatemala were analysed as before.[136] These sites were substantially closer to shore even though they were collected from deep waters around the 2000 m isobath. As a consequence of this proximity to a coastal tropical environment, the terrestrial markers are dominant and the total concentrations are substantially greater than those measured in the offshore sites. The mean profile expressed as proportions of the total concentration due to wide changes in concentration (from 337 to $3380\,ng\,g^{-1}$) can be seen in Figure 7.17.

Although a minor peak in the C_{16} can be seen in these data, the C_{22} is a considerably more important contributor to the total alcohols present. It may be concluded from these three studies that the proximity to terrestrial runoff is crucial in determining the fatty alcohol profile and not water depth. Marine production at remote oceanic sites is small compared to the terrestrial influence from continental runoff.

7.1.12   Continental Slope, Southwest of Taiwan (J1)

In a study of fatty alcohol degradation, Jeng *et al.*[117] collected a box core from the continental slope of Taiwan in 354 m of water. The sediments were sectioned every 4 cm and the fatty alcohols were extracted in a hexane–chloroform (2 : 3) solvent, partitioned into chloroform–methanol (4 : 1) and hydrolysed overnight. Fatty alcohols from C_{14} to C_{28} were reported, although the authors

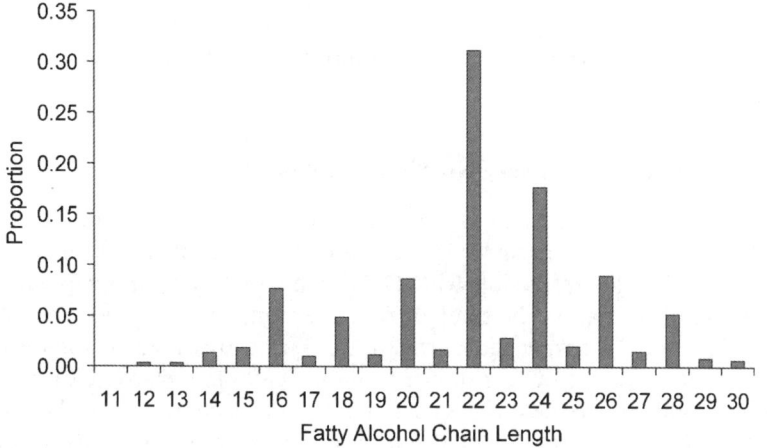

Figure 7.17 Fatty alcohol chain length profile from sediments in the Guatemalan Basin. After Howell.[136]

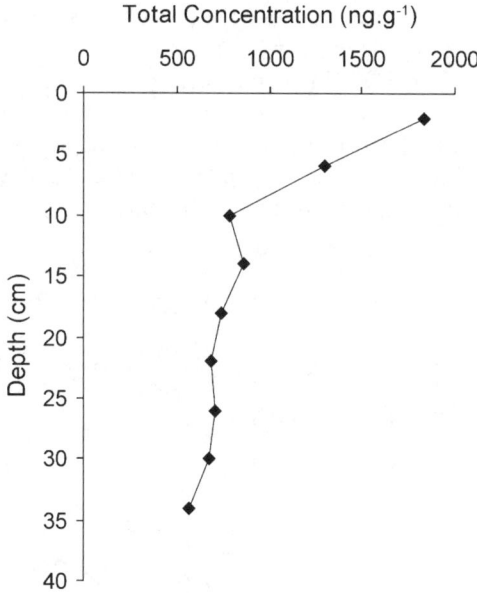

Figure 7.18 Total fatty alcohol concentration in a short core on the continental shelf off Taiwan. After Jeng *et al.*[117]

used C_{19} as an internal standard which is likely to be present in the samples naturally as well; no correction was made for this. The total concentrations in this core can be seen in Figure 7.18. High concentrations near the surface (1839 ng g^{-1} in the 0–4 cm fraction) decrease relatively rapidly to ~ 800 ng g^{-1} by 10 cm and degrade with a slower rate at deeper depths. From these data, the authors calculate the degradation rate of fatty alcohols as between 0.010 and $0.007\,y^{-1}$. These aspects are reported in more detail in Chapter 5.

7.1.13 East China Sea, North of Taiwan (J2)

In a follow-up study investigating the source of organic matter in the East China Sea, Jeng and Huh[55] collected suspended and settled sediments from a range of sites. After Soxhlet extraction with DCM–methanol (1:1) and saponification with KOH in methanol, the samples were fractionated on a silica gel column. The data from the suspended particles are partly presented in Figure 7.19a; here, the ratio of C_{24} to C_{16} is shown and it indicates the relative enrichment of the long chain compounds in the near-shore samples compared to the offshore samples. The concentrations of the fatty alcohols are significantly greater in the suspended particles compared to settled sediments and reach $166\,\mu g\,g^{-1}$ for C_{22} at site 18.

With regard to the settled (bottom) sediment, the maximum concentration reached $476\,ng\,g^{-1}$ for C_{22} at site 36, three orders of magnitude less than the suspended particulate concentration. The ratio between the terrestrial bio-marker (C_{24}) and the marine marker C_{16} is also different with only one value in this latter case less than 1.0. This is most likely due to the more rapid degra-dation of the short chain compounds in the sediment compared to the longer chain alcohols. The pattern of distribution between the two marker ratios is also different suggesting that the suspended particulate matter of the surface waters do not reflect the final deposited sediments.

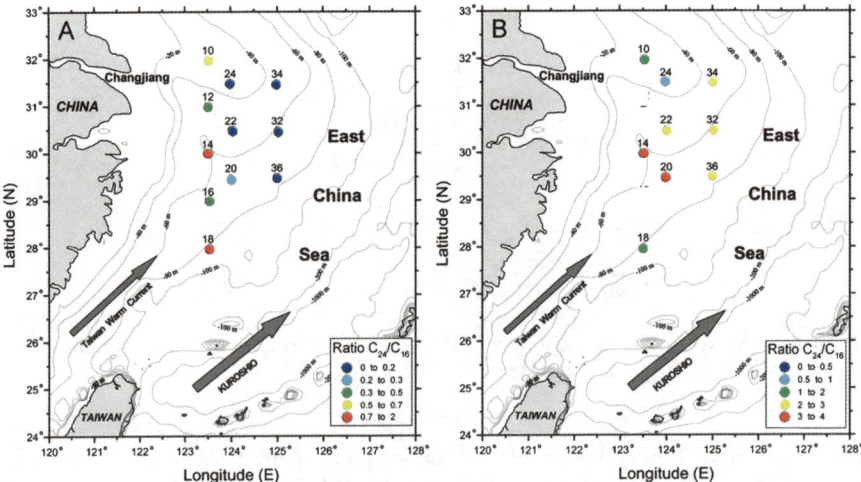

Figure 7.19 Ratio between C_{24} and C_{16} in (a) the suspended particulate matter and (b) the settled sediments collected in the East China Sea. Numbers are sample sites from Jeng and Huh.[55]

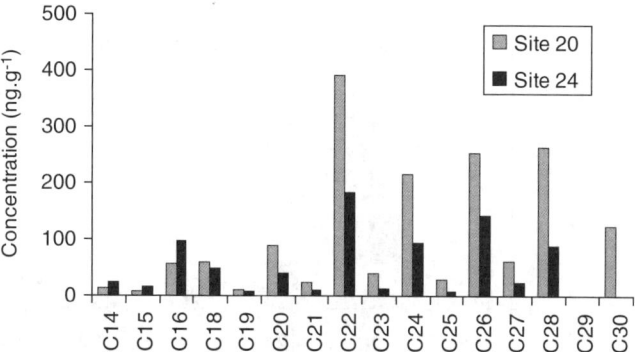

Figure 7.20 Fatty alcohol chain length profile for surface sediments from the East China Sea. Data after Jeng and Huh.[55] Site numbers refer to locations shown in Figure 7.19.

The authors of the work[55] suggest that the region is dominated by marine production of fatty alcohols, although the settled sediment fatty alcohol profiles do not reflect this. In the two extreme cases of the ratio shown in Figure 7.19b, the concentration profile indicates a substantial terrestrial component (Figure 7.20). If these data are compared to the oceanic core samples of Howell,[136] considerable differences can be seen. The Chang Jiang (Yangtze River) is the largest river in Asia and the third longest in the world. It might be expected that considerable amounts of terrestrially derived organic matter would be entering the East China Sea by this route and depositing on the continental shelf off the river mouth. This may provide the long chain fatty alcohols seen in the samples.

7.2 The Terrestrial Environment

7.2.1 Pasture Land, Southern Australia (K)

In a study of the faecal material of grazing animals in Australia, Nash *et al.*[137] quantified a few fatty alcohols in surface water runoff from pasture land. Somewhat surprisingly, the authors only report alcohols in the C_{26}–C_{32} range although this may be due to concentrating their efforts on the sterols which elute on a typical GC run in the same region. Therefore, the absence of C_{16}–C_{25} in these data does not mean they were not present; they just were not quantified.

The data for a typical sample pair are presented in Figure 7.21. These data indicate a predominance of the C_{26} fatty alcohol with smaller amounts of the other even chain length compounds; no odd chain length compounds were reported.

7.2.2 Prairie Zone Soils, Alberta, Canada (L)

Soils can act as a short term repository for many biologically derived compounds. Due to the relative stability of waxes, these may be preserved for

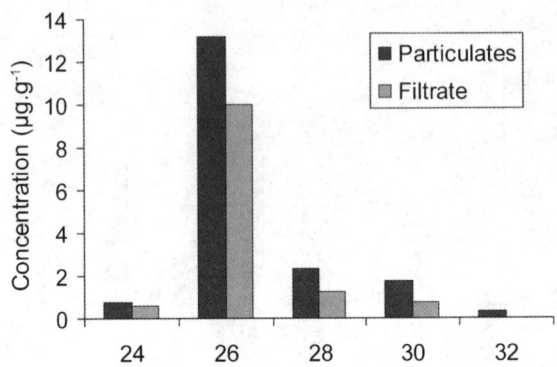

Figure 7.21   Concentration of long chain fatty alcohols by chain length in runoff from pasture land in Australia.[137] Units are $\mu g\,g^{-1}$ for the particulate fraction and $\mu g\,l^{-1}$ for the filtrate.

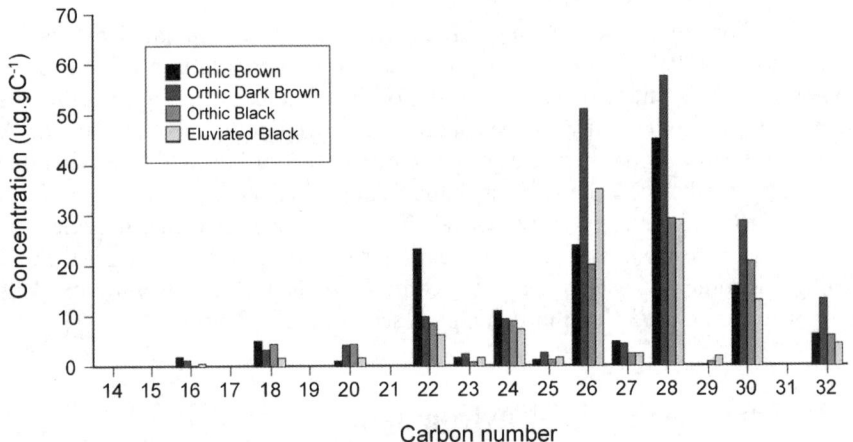

Figure 7.22   Fatty alcohols by chain length in various Canadian soils. After Otto *et al.*[138]

several years. In most cases, these are derived from terrestrial plants and so the profile is similar to that of local plant species (Figure 7.22).

In other locations, it may be that the waxes are derived from animals such as sheep.[103] In such a case, the presence of these waxes may reduce the fertility of the soil and steps may need to be taken to reduce their levels in the soil prior to cultivation.

7.3   UK Studies

As well as the globally diverse studies reviewed above, several have been undertaken in the UK. Most of these are concentrated on marine samples and also on the western side of the UK (Figure 7.2). Some of these data are taken from unpublished sources.

7.3.1 Conwy Estuary: Estuarine Core (50 cm) (1)

As part of a project investigating the use of multivariate statistics with bio-markers, Mohd Ali[110] collected a core from the Conwy estuary, North Wales. This was a muddy location 3 km from the mouth of the estuary. The system receives terrestrial organic matter from the extensive mixed forests in the catchment as well as domestic sewage and a small amount of industrial wastewater from the towns of Conwy and Llandudno. Fatty alcohols from C_{14} to C_{29} were present in the core samples after an alkaline saponification extraction procedure.[56]

Near the surface, short chain compounds (*e.g.* C_{16}) dominated, but at greater depth, longer chain compounds increased in importance (Figure 7.23). The concentrations were of the order of $ng\,g^{-1}$ DW with a maximum value of $939\,ng\,g^{-1}$ DW. The mean chain length increased from 19.5 at the surface to almost 22 by 50 cm depth.

There were several other changes in alcohol profile down the core with the percentage of branched (*iso* and *anteiso*) chain compounds reaching a subsurface maximum (11.2% at 9 cm) and decreasing to almost zero at depths > 40 cm (Figure 7.24). This subsurface maximum is a relatively common occurrence in cores and may be due to increased bacterial biomass degrading the organic matter as it settles out of the water column. The oxygen content of the sediments also decreases with depth in cores and there may be a change from aerobic to anaerobic bacterial groups which have a different fatty alcohol suite. Simple measures such as the percentage branched do not usually reflect these changes, but multivariate statistics do tend to highlight them (Chapter 9).

At the deeper depths, the samples were dominated by the longer chain compounds from terrestrial plants. These waxy compounds are generally more

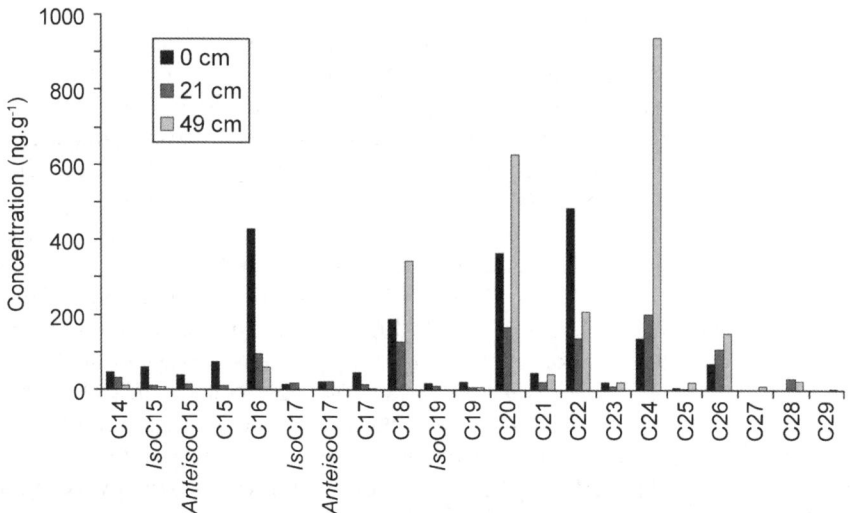

Figure 7.23 Fatty alcohol chain length profiles from a shallow (recent) core in the Conwy estuary, North Wales.

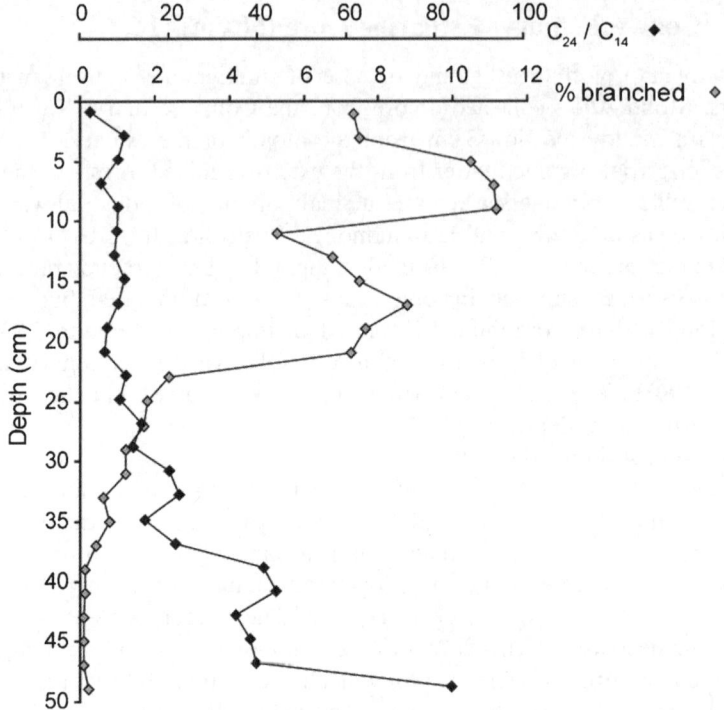

Figure 7.24   Change in percentages of branched fatty alcohols and the C_{24}/C_{14} ratios down the core from the Conwy estuary, North Wales. This may be a measure of either (i) change in source from marine to terrestrial with increasing depth or (ii) preferential degradation of the short chain alcohol.

resistant to degradation and are preserved in the core. This is unlike the observation made for alcohols in the STP effluent (Figure 5.12) where long chain compounds were metabolised or removed during treatment. The concentration per gram dry weight also increases with depth as more deposited material is compressed into a smaller section. Since these compounds are not degrading over that time scale, the concentration increases on a per gram dry weight basis.

The degradation rate may be studied in a similar manner to other sites (see Chapter 5). However, no estimate was made of the sedimentation rate in this system and so a value of $0.3\,cm\,y^{-1}$ was adopted from previous studies. A plot of the phytol degradation can be seen in Figure 7.25. The trendline provides a measure of the degradation rate ($= 0.019\,y^{-1}$) and is consistent with that measured by Jeng et al.[117] for the extractable phytol. Considerable variation may be present in this value depending on the true sedimentation rate.

Further measures of the rate of change either due to change in source or preferential degradation of the short chain compounds can be seen by using the C_{24}/C_{14} ratio (Figure 7.24). In this case, low values near the surface indicate a relatively high C_{14} composition which declines with increasing depth or age. In the deepest sample, the amount of C_{24} completely dominates the amount of C_{14}.

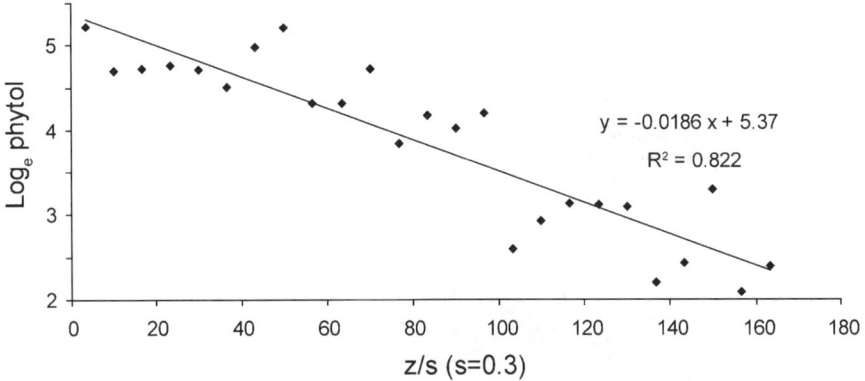

Figure 7.25 $\log_e$ of the phytol concentrations from a Conwy core. A sedimentation rate of $0.3\,\mathrm{cm\,y^{-1}}$ was assumed.

7.3.2   Mawddach Estuary: Surface Sediments (2)

Mohd Ali[110] also investigated the surface sediment distribution of a range of lipid biomarkers including fatty alcohols, fatty acids and sterols in the Mawddach estuary, North Wales. This is a relatively clean, sandy location with no industrialisation, although there are several domestic sewer outflows into the estuary. Surface scrapes of sediments were taken at the locations indicated on the map in Figure 7.26. Some samples were collected above the areas impacted by the tide to ensure a terrestrial signature was obtained.

The profile of fatty alcohols in the various samples changes in response to several environmental parameters. The principal factor is the salinity as marine samples (high salinity) should be dominated by short chain compounds and terrestrial samples (low salinity) will have long chain compounds. Samples from the middle reaches of the estuary will have a mixture of both. Examples of this can be seen in Figure 7.26. This change in signature from short chain at the mouth to long chain above the tidal limit leads to a systematic change in the mean chain length for all the *n*-alkanols in the samples (Figure 7.27). At the mouth, the mean chain length is $\sim C_{17}$, which becomes $> C_{22}$ in the rivers above the tidal limit. Intermediate values in the estuarine portion of the estuary demonstrate the mixing between the two source types. Small scale changes in the values may represent where tributaries enter the main estuary.

The differences in the profile enable signatures to be developed for the different potential sources of organic matter in the system. This will be considered in more detail in Chapter 8.

7.3.3   Menai Strait: Surface Sediments (3)

As part of an ongoing study of lipid biomarkers in the surface sediments of the Menai Strait, fatty alcohols and sterols are measured in short cores from intertidal mud annually. The results of the surface sediment alcohol profile

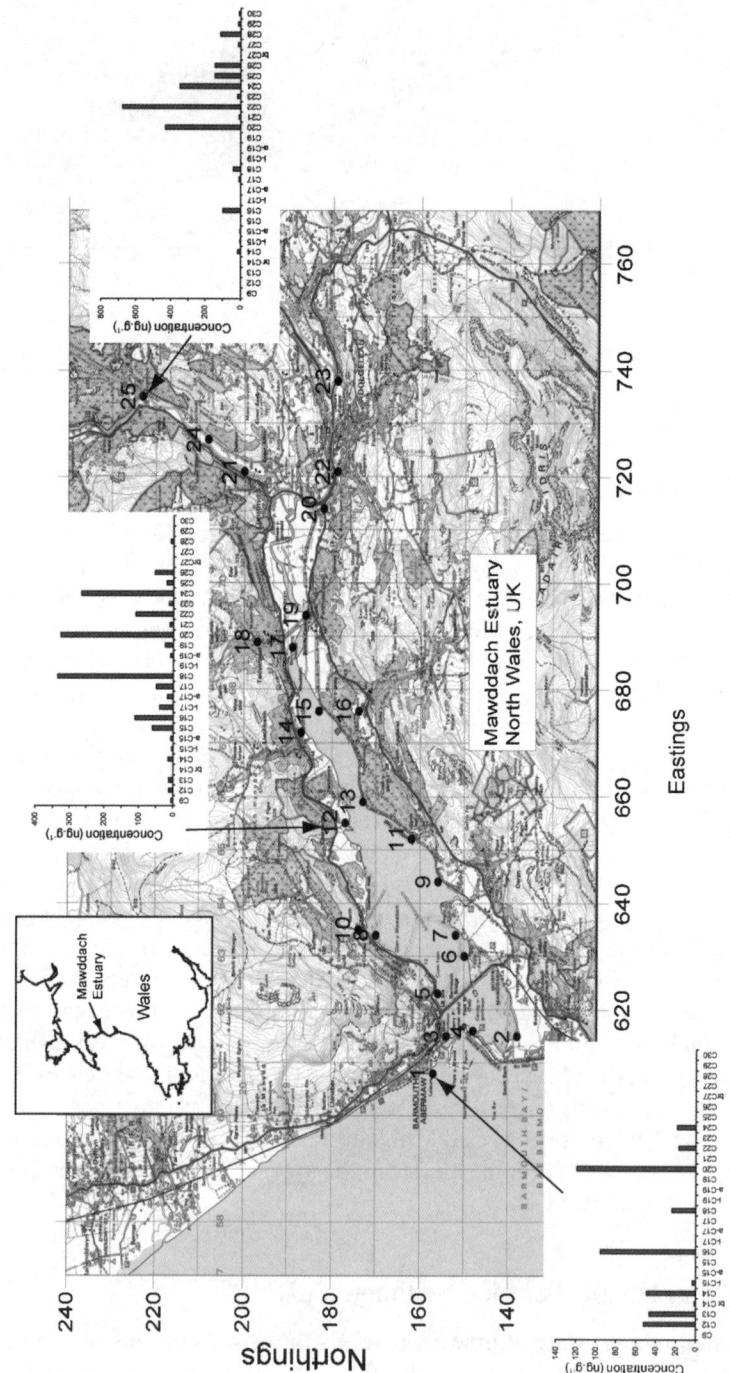

Figure 7.26 Sample location map for sites in the Mawddach estuary, UK. The sites to the east of the area are above the tidal limit. The fatty alcohol chain length profiles are shown from the marine environment (site 1), mid-estuary (site 12) and above the area impacted by tides (site 25).

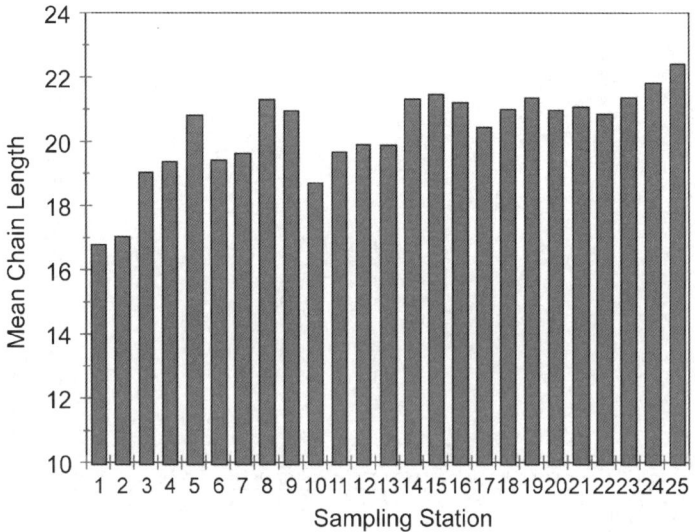

Figure 7.27 Mean chain length of the *n*-alkanols through the Mawddach estuary from the marine environment (site 1) to above the area impacted by tides (*e.g.* site 25).

extracted by alkaline saponification are shown in Figure 7.28. In all years, there is a weak bimodal distribution with peak concentrations in the C_{22}–C_{26} region representing terrestrial organic matter inputs and again in the C_{16}–C_{18} region from marine animal synthesis. The concentrations varied between years but reached a peak of $\sim 500\,ng\,g^{-1}$ wet weight for C_{22} in 2004. Branched chain compounds are also present together with odd chain length alcohols indicating the presence of bacterial biomass.

Plots of the core data (not shown) for each year show a characteristic high concentration at the surface, which decreases with depth, indicating the *in situ* degradation of these compounds by the indigenous bacterial population.

7.3.4 Loch Riddon, Scotland: Mid-length Marine Core (4)

To investigate the long term changes of lipid biomarkers,[110] a series of cores from the sea lochs surrounding the Clyde sea area were collected in 1998. Detailed analysis was conducted on a 1.5 m core from Loch Riddon (Figure 5.5), where sterols, fatty acids and fatty alcohols were measured after alkaline saponification.

The change in alcohol profile can be seen in Figure 7.29. Short chain compounds again dominate in the surface samples and decrease with depth while the longer chain compounds come to dominate at the bottom of the core. The increase in concentration of the C_{22} compounds is not as marked as for the Conwy estuary as this catchment has relatively little wooded areas and so the terrestrial component entering the system is smaller. Even so, the compounds are still preserved in the 145–150 cm section which is of the order of 1000 years before present (BP) based on the PAH profile.

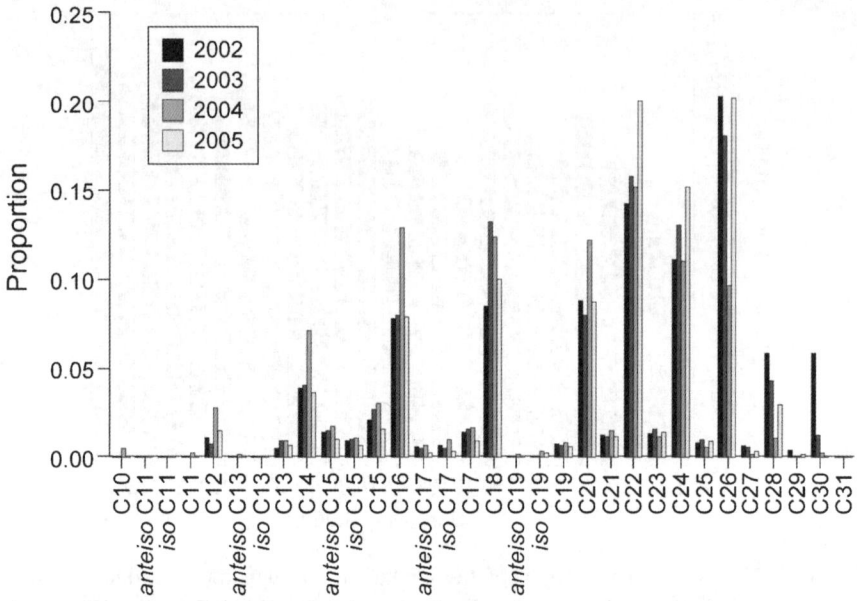

Figure 7.28 Fatty alcohol profile expressed as proportions for surface sediment in the Menai Strait collected in February 2002–2005.

It is interesting to note that the C_{14} compounds are still present in the deepest sample despite being ~ 1000 years old. The rate of degradation in this system may be less than for the Conwy estuary due to the deeper cold nature of the sample location and lack of bioturbation.

The ratio between short chain ($<C_{19}$) and long chain ($>C_{18}$) compounds can also be seen in Figure 7.29; the surface samples have values between 5 and 8 while the deeper sections are tending to a ratio of 1.0. The rate of change of this ratio tends to imply that the degradation rate is slower than that in the Conwy estuary.

7.3.5 Clyde Sea, Scotland: Surface Sediments (5)

The waters of the Clyde Sea area in Scotland (Figure 7.30) receive multiple inputs from rivers of southern Scotland that must first flow through the sea lochs; the sills marking the mouths of the lochs trap much of the sedimentary matter leaving only small amounts to be transported to the open sea. The Clyde River on which the city of Glasgow is situated receives much anthropogenic waste from industrial and domestic sources. As well as *in situ* (autochthonous) production by the marine biota, sewage wastes have been disposed of through pipes to the sea and sludge disposal sites. Grab samples were collected from the surfaces at a range of sites throughout the system during 1998.[110] All samples were extracted by alkaline saponification and fatty alcohols quantified along with fatty acids and sterols.

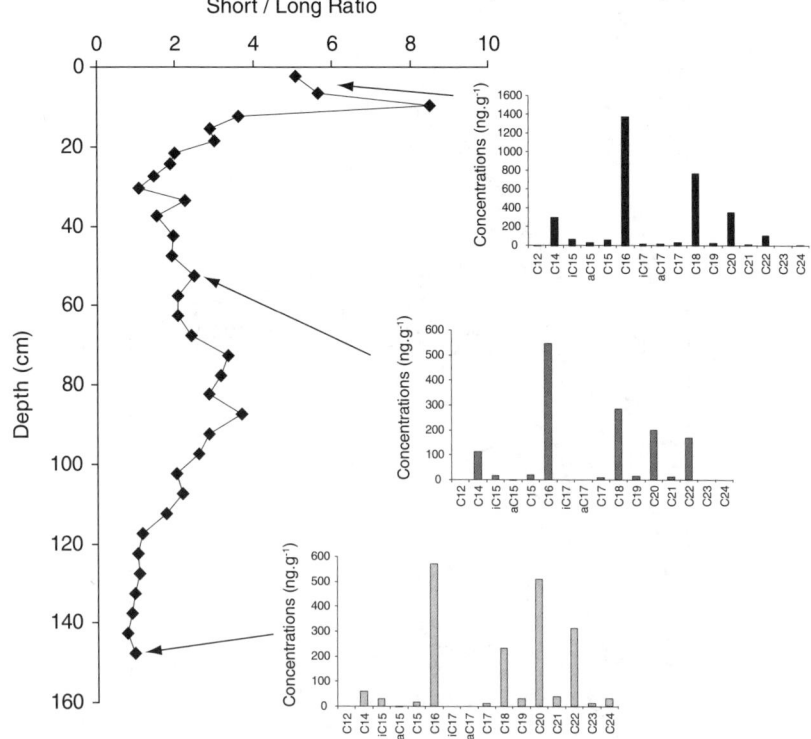

Figure 7.29 Short/long chain ratio and fatty alcohol chain length profiles for the surface, middle and bottom of a 150 cm core from Loch Riddon, Scotland. Note the change in scale for the concentrations, with ~3 times more C_{16} at the surface compared to the deeper samples.

The concentrations of fatty alcohols ranged from below detection limit for some of the short chain compounds to 2216 ng g^{-1} DW for C_{22}. An example of the distribution can be seen in Figure 7.31 which shows the concentrations of C_{16} and C_{22} fatty alcohols in the surface sediments.

The distributions are somewhat complex with no clear trend from a marine through to terrestrial gradient. This may be due to the mixing of wastewaters and domestic effluent with river water before discharge into the sea area. The sea lochs also trap many compounds from both marine and terrestrial sources and as the data from the Loch Riddon core shows, degradation may be slower here than in shallower waters. However, there are generally higher concentrations of the short chain compounds from C_{11} to C_{16} in the southern more marine areas and higher concentrations of long chain (*e.g.* C_{22}) compounds in the sea lochs (Figure 7.31). The branched chain compounds also show complex patterns and are best investigated with multivariate statistical methods, as reviewed in Chapter 9.

From a biomarker viewpoint, in this location fatty alcohols were not as useful in the tracking of organic matter as the sterols and fatty acids.[110] This is discussed further when considering PCA and PLS analysis in Chapter 9.

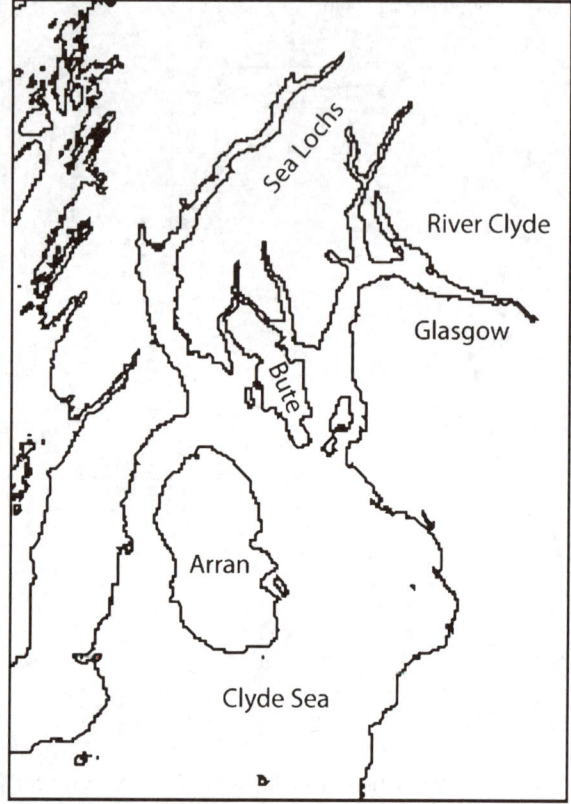

Figure 7.30   Clyde Sea area of Scotland. Major inputs come from the sea lochs to the
north and the Clyde River to the north east.

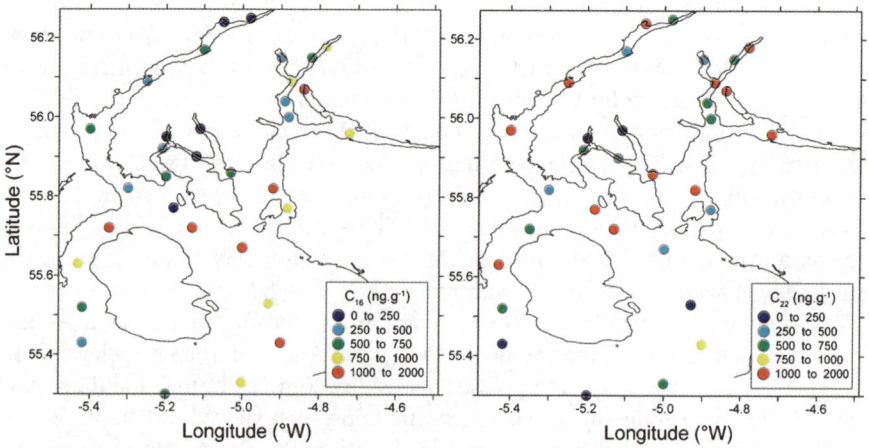

Figure 7.31   Spatial distribution of the C_{16} and C_{22} fatty alcohols in the surface
sediments of the Clyde Sea.

7.3.6 Looe Pool, Cornwall: Coastal Lake (6)

Looe Pool is a eutrophic coastal lake in southwest England. It has received nutrient inputs from agriculture in the recent past and has started to have blooms of *Hydrodictyon reticulatum*, a fast-spreading nuisance green alga.[148] A study was conducted on the sediments of the pool in which the fatty alcohols were measured.[139] The mean profile of the alcohols in the sediments can be seen in Figure 7.32. These samples were taken before the first reports of the nuisance blooms of the green alga, which became dominant in 1993,[149] although cyano-bacterial blooms had been present earlier.

The distribution of alcohols is typical of a terrestrial system with significant amounts of the C_{22}–C_{28} fatty alcohols and almost no short chain compounds present. The total concentrations reached $330\,\mu g\,g^{-1}$ DW of sediment suggesting a high loading of organic matter.[139]

7.3.7 Bolton Fell Moss, Cumbria: Mire (7)

As part of a study in an ombrotrophic ("rain nourished") mire, Avsejs[150] investigated the lipids in short cores (30 cm $\equiv$ 160 years BP) of peat. The full results are reported elsewhere.[70,151,152] The location of this peat bog is in northern England close to the Scottish border (Figure 7.2). In his work, Avsejs[150] extracted samples without alkaline saponification and so these data are only for free, extractable fatty alcohols. The fatty alcohol profiles of the samples were very similar and the mean values ($n = 7$) are presented in Figure 7.33; both C_{22} and C_{24} were present in high concentrations relative to the other compounds.

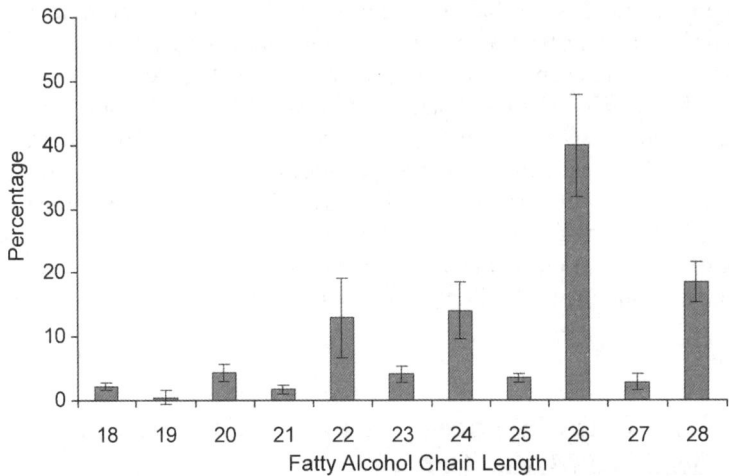

Figure 7.32 Fatty alcohol chain length profile for the surface sediment in Looe Pool, Cornwall. $N = 6$ with ± 1 standard deviation. Data after Pickering.[139]

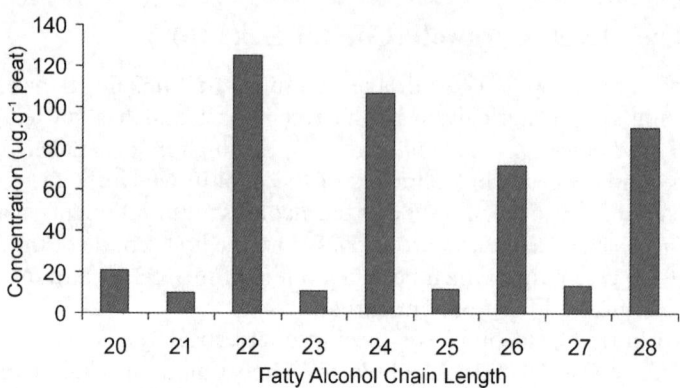

Figure 7.33   Mean concentration of fatty alcohols measured in seven peat cores. After Avsejs.[150]

The results are expressed in $\mu g\,g^{-1}$ of dry peat since the mineral content of these samples will be low. The highest value was $226\,\mu g\,g^{-1}$ of dry peat for C_{22} in a sample dominated by *Eriophorum vaginatum*.

7.3.8   Lochnagar, Scotland: Mountain Lake (8)

Lochnagar is a mountain lake 788 m above sea level in the Scottish Highlands close to Balmoral in Aberdeenshire. This system was investigated by Scott[140] as part of a study into sedimentary biomarker records of climate variability in the Holocene period. The main results are reported in Dalton *et al.*[153] A small number of fatty alcohol results are reported in a core with maximal concentrations in the C_{22} to C_{30} chain length range: the mean concentration for these long chain terrestrial plant-derived compounds was 0.3–0.4 mg g^{-1} TOC. There was a strong even over odd dominance at all chain lengths (Figure 7.34) with the shorter C_{15} and C_{17} compounds being below detection limits. Scott concludes that these compounds are principally derived from higher plant sources including peat, the degraded product of many plants in such upland areas (see Chapter 3 for profiles). In Scott's analysis of local peat, the C_{22} fatty alcohol was the most abundant compound.[140] Scott attributes the relative absence of short chain fatty alcohols in the core to a small *in situ* production by algae and dominance of allochthonous (terrestrial higher plants) organic matter.

7.3.9   Loch Lochy, Scotland: Freshwater Core (9a); Loch Eil, Scotland: Marine Core (9b)

Cores were collected in freshwater and saltwater lochs that were only 16 km apart. The freshwater core was retrieved from 116 m water depth and the

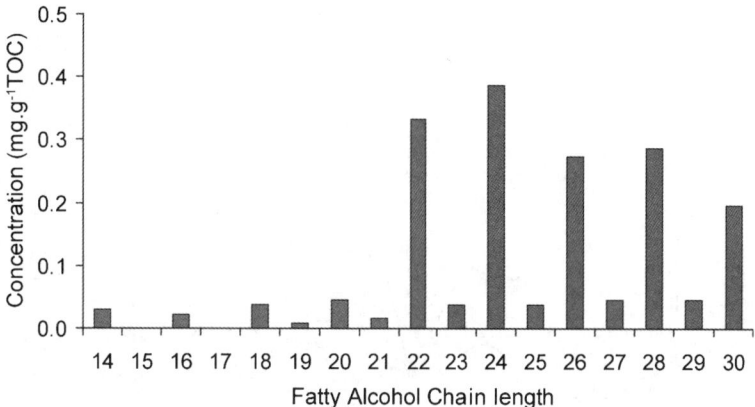

Figure 7.34 Mean fatty alcohol distribution in a sediment core from Lochnagar, Scotland. After Scott.[140]

saltwater one from 70 m depth. The two systems are loosely connected by the Caledonian Canal and there is a small water flow from the terrestrial system to the marine one. Both cores were analysed for a range of chemical and physicochemical parameters including lipids which were reported by Hotham.[109]

Overall, the concentrations in both cores were significantly lower than those found in the shallow water cores probably due to increased degradation during settlement of the organic matter through the water column and lower production/input. This is consistent with the results from Wakeham *et al.*[114] shown in Figure 5.9.

One of the interesting features of the data is the age of some of the compounds found; the freshwater core base was tentatively dated to be around 8000 years BP due to encountering the post-glacial clay layer typical of this area. Most compounds were quantifiable even at this age/depth, although the concentrations were very small (*e.g.* 0.0627 ng g^{-1} for C_{12}). The seawater core was younger at an equivalent depth due to higher deposition rates. Near-surface concentrations in the seawater core were less than in the freshwater core, although at greater depths they were similar (Figure 7.35a and b).

The branched chain compounds in both cores had an interesting profile (Figure 7.35c); due to lower deposition rate in the freshwater core, no subsurface peak in values could be seen. It could, however, be seen in the seawater core. The trend also showed increasing percentages with increasing depth which may appear counterintuitive. This is likely to be due to the slower degradation rate of the branched compounds compared to their straight chain equivalents. Therefore, the net effect is for the percentage branched to increase. Inspection of the concentration data does show that in the freshwater core the concentrations decrease with increasing depth. This is less clear in the seawater core where increased concentrations are seen in the last two samples. The reason for this is not clear.

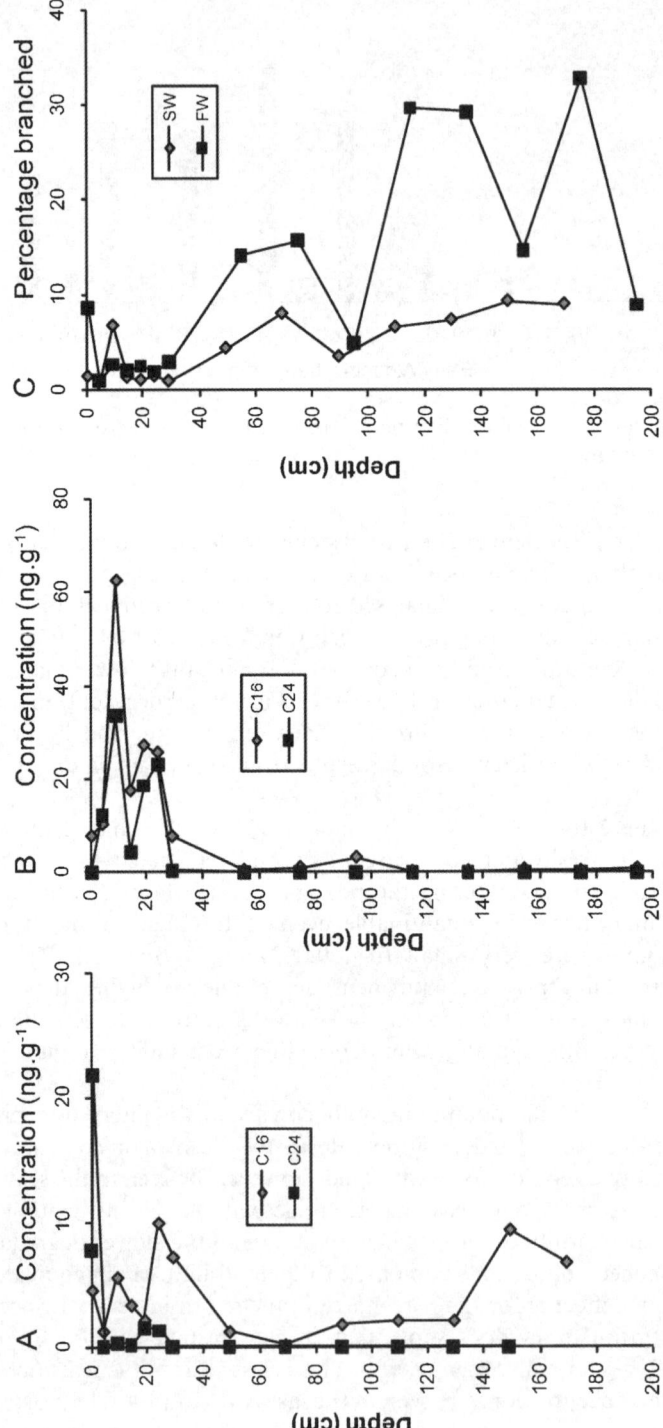

Figure 7.35 Concentrations of a short chain (C_{16}) and long chain (C_{24}) fatty alcohol in (a) a seawater and (b) freshwater core. (c) Percentage of branched compounds in these two cores (SW, seawater core; FW, freshwater core). It should be noted that although the core lengths are approximately the same, they represent very different depositional times.[109]

Table 7.3 The maximum concentrations for C_{16} and $C_{22/24}$ fatty alcohols measured at a range of locations around the world. Concentrations are per gram dry weight.

Environment	Maximum C_{16} concentrations[a]	Maximum $C_{22/24}$ concentrations	Number of studies
Coastal marine sediments	402–1961 ng g^{-1}	112–5500 ng g^{-1}	7
Estuaries and lagoons	1384–8890 ng g^{-1}	27–818 ng g^{-1}	8
Ocean sediments	12–404 ng g^{-1}	635–1000 ng g^{-1}	6
Suspended sediments	ND–2737 µg g^{-1}	166–1847 µg g^{-1}	2
Freshwater	ND	740 ng g^{-1}	2
Lake sediments	63 ng g^{-1}	98 ng g^{-1}	1
Upland soils	ND	125 µg g^{-1}	1

[a]ND, not determined.

7.4 Environmental Ranges

The data from all samples reviewed are summarised in Table 7.3. The environment is divided into regions and the highest concentrations can be seen in suspended sediments. This is not unexpected as they are generally organic rich with a low mineral content. For other locations, areas of high productivity such as estuaries and lagoons have higher concentrations than coastal sediments which are in turn greater than offshore ocean locations.

Summary

- The concentrations of fatty alcohols measured across many parts of the world are surprisingly similar (Table 7.3).
- In most marine cases the C_{16} straight chain compounds dominate with secondary maxima in the C_{22} to C_{24} region depending on the proximity to terrestrial sources of organic matter. The C_{16} concentration may be expected to be high as it is the end of the initial fatty acid synthesis pathway from which many fatty alcohols are made.
- The concentrations in suspended matter are significantly greater than those in settled sediment as the organic matter content is usually higher. These materials may also have undergone less bacterial degradation and be more representative of local sources.
- The concentrations are greater in the near-shore and estuarine/lagoonal regions due to the high *in situ* productivity and the presence of terrestrial plant waxes. This is reflected in the high C_{22} or C_{24} values. However, due to the preservation of the long chain moieties, these compounds may be found at locations remote to their origin, both in time (core depth) and space.

CHAPTER 8
Using Fatty Alcohols as Biomarkers

How can fatty alcohols be used to indicate the source of organic matter in the environment?

Fatty alcohols may be used as biomarkers to indicate the source of organic matter in the environment.[56] The range of different alcohol profiles shown in Chapter 3 indicates the ways in which these compounds may be used. The major biomarker indicators are:

1. Bacterial biomass on a relative scale by use of the odd/even ratio (carbon preference index) and the percentage branched chain compounds present.
2. Marine faunal presence through use of the short chain even carbon compounds.
3. Terrestrial plant (and animal) input through use of the long chain even carbon compounds.
4. Photosynthetic activity through use of the phytol component derived from chlorophyll.

8.1 Stable Isotopes

More sophisticated analyses using stable isotopes of carbon ($\delta^{13}C$) and hydrogen (δ^2H) may enhance the discrimination between compounds and profiles in complex multi-source environments.[154] As well as containing ^{12}C and ^1H in the fatty alcohol molecule, there will also be a small percentage of naturally derived ^{13}C and ^2H (D). The compositional differences with the main isotopes may change as chemical or biochemical processes may act preferentially on one isotope or the other. These differences can be expressed as the $\delta^{13}C$ and δ^2H values referenced to a standard, usually Pee Dee Belemnite for

Fatty Alcohols: Anthropogenic and Natural Occurrence in the Environment
By Stephen M Mudge, Scott E Belanger, and Allen M Nielsen
© Copyright 2008 ERASM (the joint surfactant environmental research platform of AISE and CESIO) and SDA
Published by the Royal Society of Chemistry, www.rsc.org

carbon and Standard Marine Ocean Water for hydrogen. The δ value is calculated from the following equation:

$$\delta^{13}C = \left(\frac{R_{\text{sample}}}{R_{\text{standard}}} - 1 \right) \times 1000$$

where R is the ratio $^{13}C/^{12}C$ and the standard is Pee Dee Belemnite. When calculating values for hydrogen, Standard Marine Ocean Water is used as the standard. More background on this methodology can be found in Philp and Kuder.[155]

Since fatty alcohols come from a wide range of potential sources (*e.g.* C_{16} from animals, plants and anthropogenic synthesis from oil), the stable isotopes may offer the only unambiguous fingerprint for their sources. However, in some consumer formulations incorporating fatty alcohols, the source of the fatty acids converted to alcohols is from palm and coconut oils and, therefore, the isotopic signature will be that of those plants.

8.1.1 ^{13}C Composition

In general, marine animals and algae typically produce fatty acids and alcohols with $\delta^{13}C$ values around -20 to $-25‰$ compared to C_3 terrestrial plants which produce values closer to $-30‰$.[156] C_4 terrestrial plants such as maize (corn) tend to have $\delta^{13}C$ values close to $-12‰$.[157] Since an organism's $\delta^{13}C$ value will usually represent its diet within $1‰$,[156] a clear distinction between these different sources should be apparent. However, humans with a mixed diet will have a signature dependent upon the mix of food eaten including fish, C_3 and C_4 plants and the animals that have eaten these plants.

8.1.2 ^{2}H Composition

The range of partitioning for the deuterium atom relative to the hydrogen atom is much greater than for the carbon isotopes and values in aquatic ecosystems vary between -300 and $-100‰$ depending on the source.[158]

Recent analyses (in preparation) have explored the use of two-dimensional stable isotope analysis (^{13}C and ^{2}H) to determine the contributions that natural and anthropogenic sources of fatty alcohols make to the environment. Samples were collected from a sewage treatment plant and from potential local sources including soils, terrestrial plants, food waste and faecal matter. The stable isotope signatures were compared to anthropogenic fatty alcohol raw materials used in a range of formulations (see Figure 4.15). The results (Figure 8.1) clearly separate those raw materials derived from petroleum (A and B) and those derived from plant materials (C and D) such as palm oils. The separation between products is apparent in both the ^{13}C and ^{2}H values and within a single product (*e.g.* raw material A) as the different chain lengths have slightly

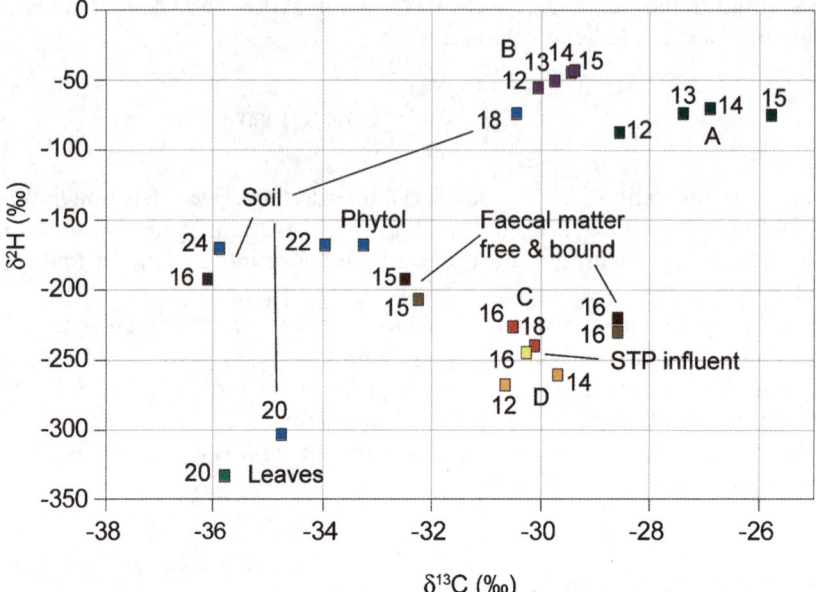

Figure 8.1   Two-dimensional stable isotope analyses of anthropogenic raw materials (A to D corresponding to products in Figure 4.15) together with a range of environmental samples. Individual labels refer to the carbon chain length. The fatty alcohols were analysed as their TMS ethers but the contribution from the TMS group has been removed in the final values.

different values. These differences are relatively small. However, compared to the differences between raw materials, the C_{16} fatty alcohol in the STP influent has an isotopic signature implying it may be a mixture of faecal matter, food wastes, naturally derived detergents and plant matter. The contribution from each potential source would require further analyses.

8.2   Bacterial Biomass

In Chapter 3 it was stated that "branched chain fatty acids in particular have been considered to reflect a bacterial origin"[43] and that both *iso* and *anteiso* branched chain acids are common to many bacteria.[44-46] In many *Bacillus* species they account for up to 60% of the total fatty acids.[13] Due to the synthetic pathway for fatty alcohols (FAR), the fatty acids should act as a good indicator of the likely fatty alcohols found in bacteria. Unlike plants, in bacteria the fatty acids are part of the cell membrane but their role is not known.

Since bacteria produce odd chain compounds while most other biota produce even carbon number compounds, the presence of odd chain compounds in samples can indicate bacterial biomass. One approach is to use the carbon preference index (CPI) as a relative measure.[55] The CPI is calculated from the following equation:

$$CPI = \frac{\sum odd\ chain}{\sum even\ chain}$$

In samples where there are numerous bacteria, it might be expected that the CPI will have a higher value than in samples where the bacterial activity and, therefore, bacterial biomass is low. There are no absolutes in this measure as there are with some other biomarkers but in an environmental situation, it may be used to indicate samples with greater bacterial activity. It is also possible to restrict the chain lengths in the calculation and confine the ratio to the short or long chain moieties.

An example of the use of the CPI can be seen in surface sediments of the Clyde Sea, Scotland (Figure 8.2). In this case, high CPI values were measured to the northwest of the Isle of Bute: these sediments were fine grained and may

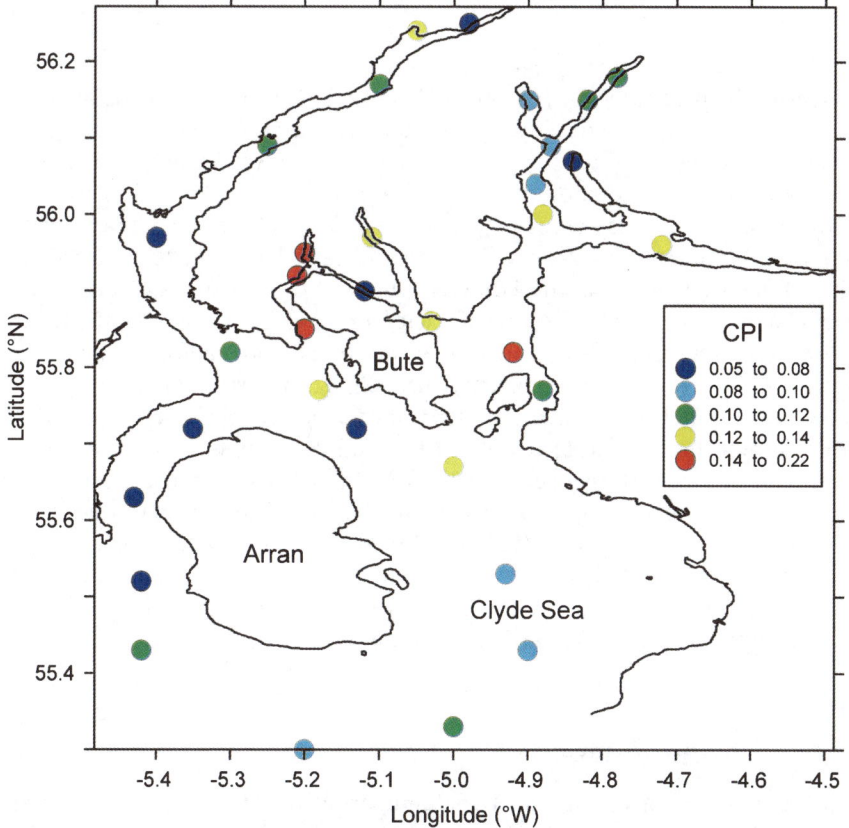

Figure 8.2 CPI in surface sediment samples across the Clyde Sea and into the Scottish sea lochs.

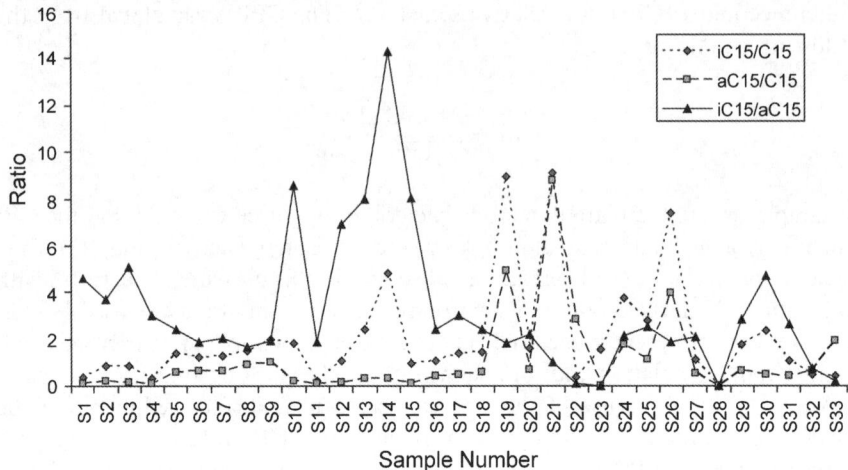

Figure 8.3   Change in preference for the three 15-carbon fatty alcohols in samples from the Clyde Sea.

be rich in marine bacteria degrading organic matter from a range of sources. Values in the next category (yellow in Figure 8.2) appear to be associated with material emerging from the Clyde River next to Glasgow. The open sea and sea lochs themselves do not seem to have such high bacterial markers. It may be theorised that these compounds are derived in part from domestic sewage.

Bacteria also produce branched chain fatty alcohols and the relative proportion of these present in a sample may also indicate the degree of bacterial activity. Different bacteria produce different suites of fatty alcohols and the profile may change according to the type of bacteria present. For example, the *iso* and *anteiso* C_{15} fatty alcohols appear to have a different association to the *iso* and *anteiso* C_{17} fatty alcohols and may also change in predominance across a sampling region. For example, data from the Clyde Sea area collected by Mohd Ali[110] shows that for the 15-carbon fatty alcohols, there is a spatial change in dominance of the three isomers (Figure 8.3). In most cases, this change in fatty alcohol signature is best approached through multivariate analyses such as principal component analysis (PCA) and further examples are given in Chapter 9. These changes probably represent changes in the bacterial assemblage which will change according to availability of oxygen, organic matter source and specific inputs from the terrestrial environment.

8.3   Marine Fauna

Since marine animals do not have the same problems of desiccation that many terrestrial organisms have, they do not require long chain wax esters on their outer surfaces (Chapter 3). The fatty alcohol profiles of marine animals tend to be rich in the short chain C_{14} to C_{16} moieties. This can be seen in several potential measures: the mean chain length will vary according to source (see Figure 7.27 for changes from $\sim C_{16}$ in marine samples which increases to $> C_{22}$ as sampling sites move through an estuarine system into areas principally impacted by terrestrial sources); the ratio of the short chain unbranched compounds to long chain unbranched compounds (or its inverse depending on whether the ratio is showing marine input to a terrestrial system or *vice versa*). The cutoff between long and short is usually determined within the dataset by multivariate analyses; those compounds that are derived from terrestrial plants cluster together and those from marine fauna tend to cluster in a different place within a loadings plot. The division may vary but tends to be in the C_{18} region. An example of these ratios can be seen in Figure 7.9b for the Ria Formosa lagoon in Portugal.

8.4 Terrestrial Plants

As outlined in Chapter 3, one of the key problems facing terrestrial organisms is the loss of water through desiccation. This has led to the prevalence of long chain fatty alcohol–acid wax esters which upon saponification provide a fatty alcohol profile rich in C_{22}–C_{30}. Several examples of these sources can be seen in Chapter 3 and again in the environmental distributions in Chapter 7. This information can be used with the above data on marine faunal source as a long/ short or short/long ratio, although the differential degradation rates must be borne in mind when considering core samples. Apparent changes in source from marine fauna at the surface to terrestrial organic matter in the deeper records may be due to faster degradation rates for the short chain compounds compared to the long chain compounds. In such cases, secondary confirmation may be appropriate with other biomarkers such as β-sitosterol and its ratio with cholesterol (see Figure 7.6).

8.5 Photosynthetic Activity

The presence of the phytol molecule in samples principally indicates chlorophyll, although it may have come from terrestrial plants and be washed into marine samples or it might represent marine production either in the water column or benthic algae. Surface sediment samples analysed for fatty alcohols frequently have high concentrations of phytol (S.M. Mudge, unpublished work) and these may be derived from (a) *in situ* benthic production, typically by diatoms, (b) pelagic phytoplankton that have died and settled out or faecal pellets after consumption by zooplankton and (c) terrestrial plant organic matter that has settled out from allochthonous detritus and incorporated into the inorganic grains. The most appropriate method for elucidating the source of phytol in marine sediments is likely to be through use of stable isotopes as the terrestrial source will have a statistically significant difference from the marine production.

Summary

Fatty alcohols may be used as biomarkers to indicate the source of organic matter in the environment. The principal approaches are:

- Odd chain length and branched compounds are diagnostic of bacteria.
- Short chain compounds can be used as biomarkers for marine fauna.
- Long chain compounds are derived from terrestrial plants.
- Phytol from chlorophyll either of marine or terrestrial origin.
- Stable isotopes of carbon (^{13}C) and hydrogen (^{2}H) in compound-specific analyses to isolate compounds with mixed sources. This may be the most appropriate long term approach in complex environments such as sewage works and their receiving waters.

CHAPTER 9
Multivariate Statistics

What extra information can be gained from these multivariate statistics methods?
Partial least squares can give quantitative estimates of each source input.

9.1 Chemometric Methods of Use With Fatty Alcohols

There are several multivariate statistical methods that may be of use when
interpreting large datasets containing fatty alcohol data. These include:

1. Principal component analysis (PCA)
2. Partial least squares (PLS)
3. Cluster analysis
4. Factor analysis
5. Multi-dimensional scaling (MDS).

The advantage of each of these methods is the ability to take all the compounds and sample observations at once and determine structure within the
data. These methods are now relatively widely used in environmental biomarker analysis and may be of use with fatty alcohols alone or as part of a
wider dataset including other compounds.

9.1.1 Principal Component Analysis (PCA)

PCA of environmental fatty alcohol concentrations or proportions may be
expected to identify those compounds that co-vary. In a two-dimensional plot
of the loadings using the first two components (PC1 and PC2), those compounds that have the same source or behaviour will aggregate together. An
example of this can be seen in Figure 9.1; these data are proportion data from
samples collected on Blackpool Beach, northwest England, during a 13-week
period including part of the designated bathing season (European Bathing

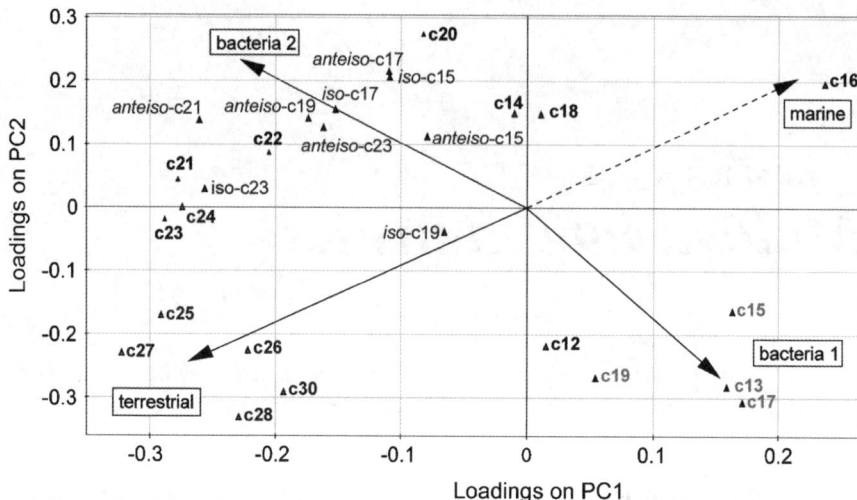

Figure 9.1 Two-dimensional loadings plot for fatty alcohols from Blackpool Beach. Four potential source aggregations can be seen: terrestrial dominated by the long chain odd and even carbon compounds; a weak marine vector generally positive on PC1 opposite terrestrial; and two bacterial vectors, 1 and 2, characterised by either odd carbon chain length compounds or branched compounds.

Waters Directive, 76/160/EEC). Eleven locations were measured on each occasion.[120] Potential source materials of faecal matter to the beach were also measured including faeces from cows, donkeys and sheep; influent and effluent to the major sewage treatment plant and surface water drains.

The major axis (PC1 explaining 19.2% of the variance in the data) appears to be a marine–terrestrial axis. However, the aggregation of the potential marine animal fatty alcohols is rather weak and is shown with a dashed arrow. PC2 (12.3%) divides two groups of compounds that are usually classified as bacterial in origin. The short chain odd carbon compounds load negatively on PC2 and positively on PC1 (showing a marine bacterial source?) while all the branched compounds are to be found in the opposite quadrant (terrestrial source?).

These putative source vectors can now be laid on the sample location aggregations: the scores plot (Figure 9.2). The potential sources to this region (colour coded in the figure) show that the influent material to the STP (Cmarin1–3), Cows and Sheep3 fall along the terrestrial vector determined from the loadings plot. In the lower right quadrant, the samples are principally effluent from the STP (Cmaref1–3), river samples (RIB1–3) and surface water drains. The loadings plot shows these samples to be enriched in the short chain odd carbon fatty alcohols. In comparison, the data in the upper left quadrant designated bacteria 2 due to the presence of *iso* and *anteiso* branched fatty alcohols are many of the environmental samples and none of the potential sources. The general location of Blackpool can be seen in Figure 9.3. These

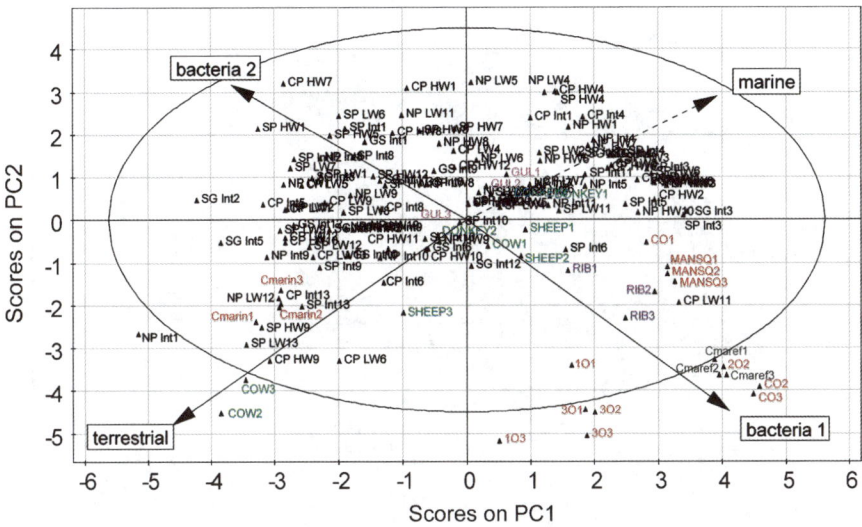

Figure 9.2 Scores plot for the data from Blackpool Beach with the putative sources from the loadings plot overlaid. The coloured sample names indicate the triplicate analyses of potential source material to the area.

aggregations suggest that the environmental samples in the upper left quadrant of Figure 9.2 have undergone a degree of degradation and these branched chain compounds are from environmental bacteria assemblages.

Another example of compound aggregation can be seen in a set of core data from Loch Riddon in Scotland (see Figure 5.5 for a location map).[110] As part of the pre-treatment of the data, all values were converted to proportions to remove the concentration effect and used without further transformation. The loadings plot (Figure 9.4) shows the main axis (PC1, 45.6%) is a marine–terrestrial axis with the short chain compounds to the right and longer chain moieties to the left of the figure. The second PC (18.1%) separates bacterial related (branched and odd chain length) compounds giving rise to two populations as before. In this case, as the data are from a core, it is possible to see a sequence through time. The scores plot (Figure 9.5) shows a sequence with depth indicated by the line. The top samples are relatively rich in short chain alcohols, C_{12} and C_{14}, together with phytol, a chlorophyll marker, and 20:1 the storage lipid prominent in copepods, the main organism in the zooplankton. This is consistent with relatively fresh marine inputs. Samples immediately below the surface have a trend towards increasing *anteiso* C_{15}, a bacterial marker. It is likely that this reflects the microbiological community metabolising the organic matter as it enters the sediments.

As the depth increases, the sequence moves towards the bottom of the scores plot towards a second set of bacterial markers, *iso* C_{15} and straight chain C_{15}. Interestingly, the C_{16} is also part of this group. This may be a bacterial group more tolerant of lower oxygen tensions, potentially even anaerobic

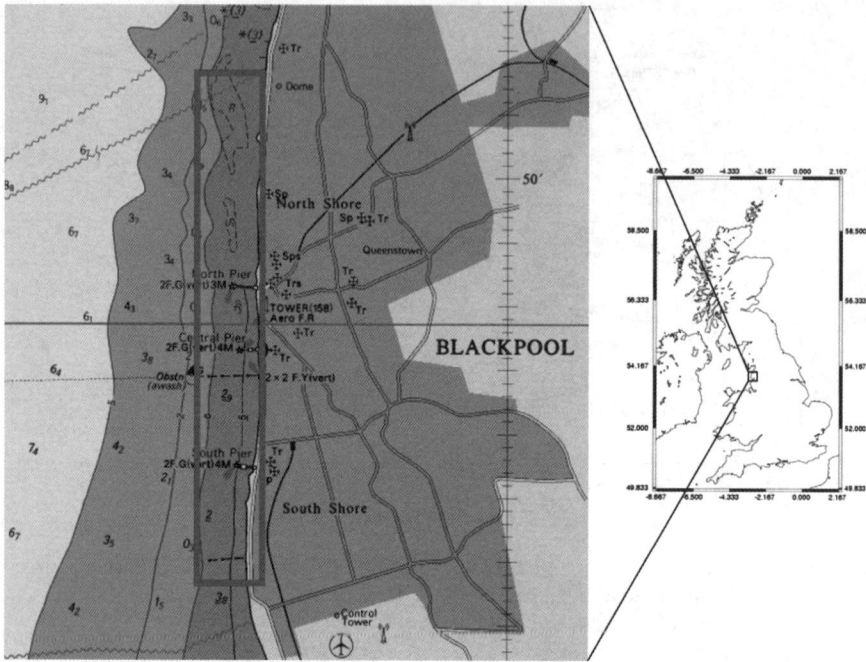

Figure 9.3   Location map for the samples collected on Blackpool Beach, with the sampling zone in the intertidal outlined by the box.

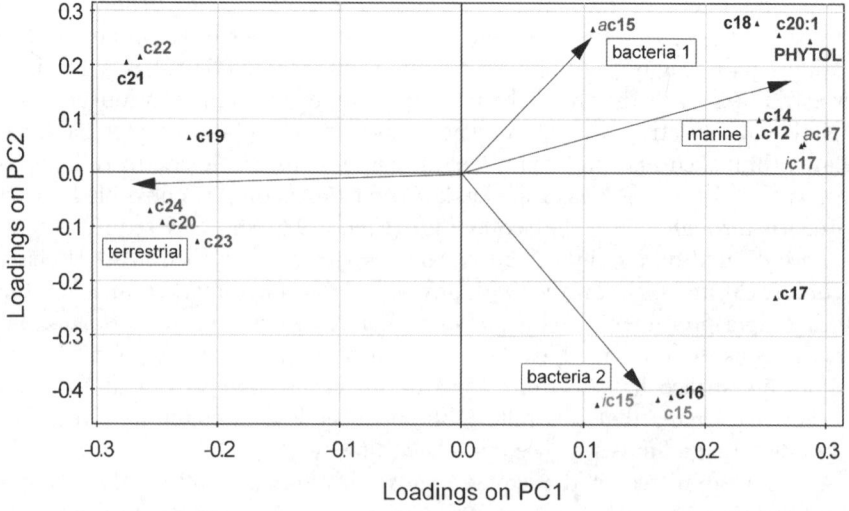

Figure 9.4   Loadings plot for fatty alcohols from a 1.5 m core collected from Loch Riddon, Scotland.

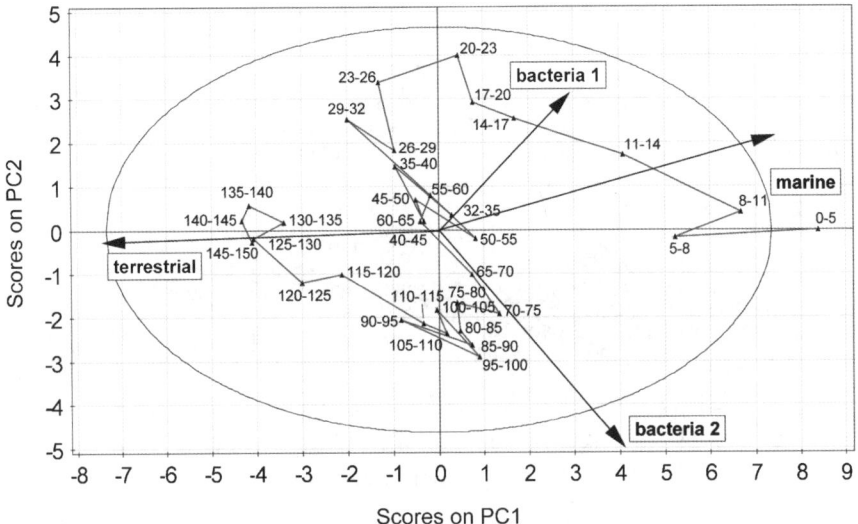

Figure 9.5 Scores plot for samples down a core. The labels refer to the sampling depth (in cm). The top of the core is to the right and the bottom to the left.

communities, although the redox potential was not measured at the time of sampling. At depths greater than 100 cm, the samples tend towards the left of the scores plot and become enriched in the longer chain alcohols, both odd and even chain lengths. These are characteristic of terrestrial environments.

It is possible to explain the trends in the data as changes in source–samples near the surface that are dominated by marine inputs which have replaced older deposits of principally terrestrial origin. However, it is also possible to interpret the data in terms of differential degradation rates; the waxy long chain compounds are more stable and less readily degraded in the marine environment compared to short chain (marine derived) compounds. These data alone do not enable a definite answer to be obtained although the presence of a suite of bacterial biomarkers suggests activity along these lines. However, other biomarkers are more resistant to degradation and may be used to provide secondary evidence. A key terrestrial compound that is relatively stable is β-sitosterol (24-ethyl cholesterol) formed in the secondary thickening of higher plants.[56] A good marker is the ratio of this compound to cholesterol which is commonly thought of as being a marine marker.[56] A cross plot using the sterol ratio against the fatty alcohol ratio can be seen in Figure 9.6; different regions can be seen corresponding to different depths in the core. Three regions can be identified on the plot: A – surface samples where the rates of change of the two markers are essentially the same; B – a region where the sterol marker changes while the fatty alcohol one does not (this coincides with the change from bacterial group 1 to 2 on the PCA scores plot); C – a region where the markers agree but the rate of change is less in the alcohol marker than might be expected from the sterol data.

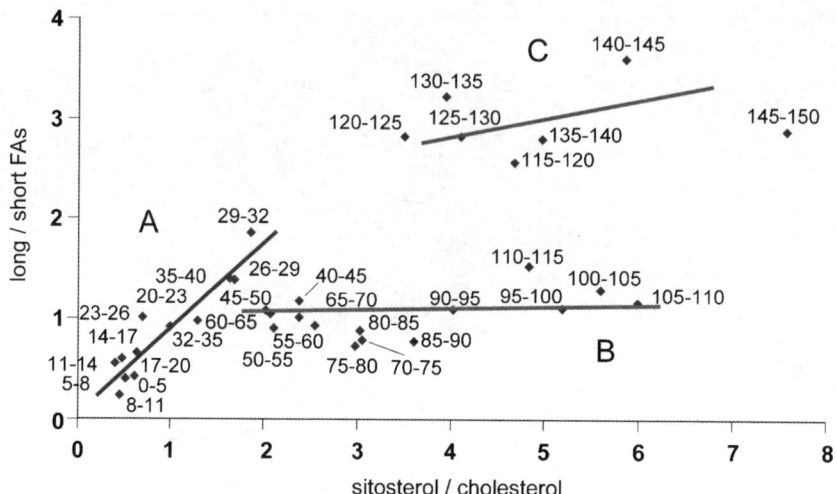

Figure 9.6   Relatively resistant sterol marker for terrestrial plant matter (sitosterol/cholesterol) plotted against the straight chain fatty alcohol marker ($\sum(C_{19}-C_{24})/\sum(C_{12}-C_{18})$). See text for explanation of the region codes.

The PCA in these two examples demonstrates the usefulness of these statistical methods in identifying the relationships between both compounds and sample sites. These data do not appear to show any enhanced concentrations of fatty alcohols that may be derived from anthropogenic usage; if there were high concentrations of detergent-based alcohols (alcohol ethoxylates, alcohol sulfates or alcohol ethoxysulfates) entering the system through the wastewater system, there would be more short chain compounds ($C_{12}-C_{16}+C_{18}$) in the samples. Figure 5.12 (the concentrations plot from the Fleetwood STP) does show increased prevalence of the odd chain fatty alcohols C_{13} to C_{19} but this includes the C_{17} which is specifically absent in the anthropogenic formulations (see Table 4.3) and also does not have the even chain components which are located elsewhere in the figure. It may be concluded from this information that these compounds are derived from biological sources rather than anthropogenic ones.

Similarly, in the core data, the surface samples are enriched in what superficially may be considered to be anthropogenic derived compounds but this group also includes phytol derived from chlorophyll and 20:1, the storage lipid from many zooplankton. Again, it must be concluded from these data that the fatty alcohols are essentially natural in origin in these two systems.

Another feature of PCA is the ability to look at the contribution each compound makes towards the overall "signature"; this can be seen in contributions plots. An example of this for the effluent from the STP at Blackpool Beach can be seen in Figure 9.7. If this effluent was enriched in fatty alcohols derived from anthropogenic sources, it might be expected to see high contributions from the $C_{12}-C_{16}+C_{18}$ compounds; in this case, the C_{17} is most prominent and the C_{16} and C_{18} are depleted relative to the average projection.

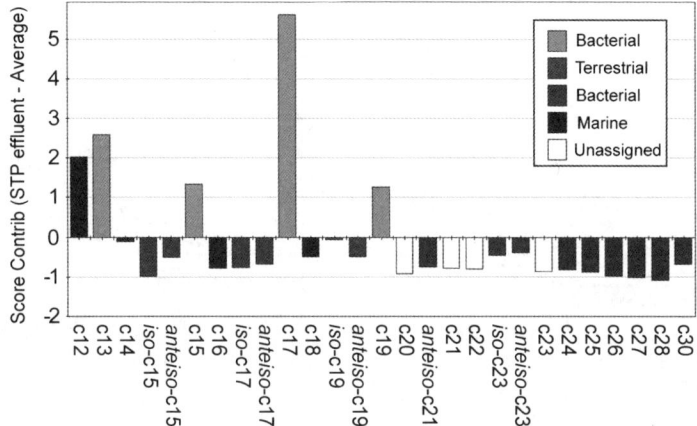

Figure 9.7 Contributions plot for STP effluent relative to the average loadings. Two groups of bacteria were apparent in this sample coincident with the groupings seen in the environment. This particular sample was enriched in odd carbon number, straight chain fatty alcohols, especially C_{17}.

The internal relationship between each of the compounds in the sample allows a signature to be developed from the chemical contribution. This can be achieved in a semi-quantitative manner using the statistical technique of PLS (see below).

A number of methods are available to improve discrimination in PCA including the use of proportion data (removes the concentration effect), log transformation (to improve normality) and addition of small values (to remove zeros by adding 50% of the limit of detection – useful if doing log transforms).[159] An example of such improvements can be seen in Figure 9.8. The loadings for fatty alcohols from the Ria Formosa lagoon can be seen in Figure 9.8a after a log transformation and the corresponding scores plot is in Figure 9.8b. These figures show a clear separation of compounds according to source with short chain alcohols to the right indicating marine materials and long chain alcohols to the left indicating terrestrial plants. The branched chain compounds show a range of associations indicating potential origins or different environments.

The locations in the scores plot (Figure 9.8b) also clearly separate according to their chemical composition and indicate the marine influenced locations compared to the terrestrial ones. Those sites adjacent to the sewage outfalls (*e.g.* 32–34) are located in the area associated with branched compounds in Figure 9.8a.

9.1.2 Partial Least Squares (PLS)

The PLS technique was developed by Wold *et al.*[160] and has evolved into a powerful tool in environmental forensics.[133,161] In essence, PLS performs PCA

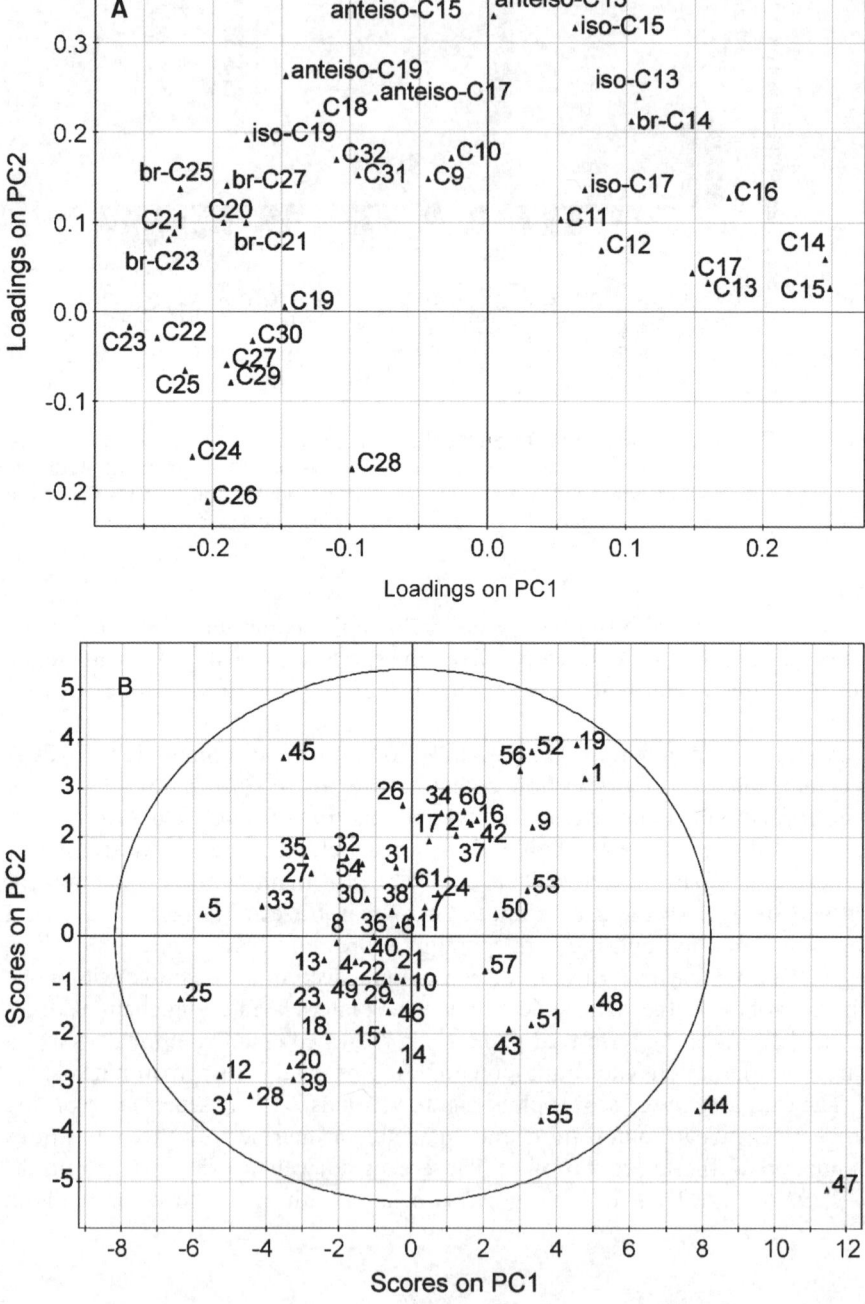

Figure 9.8 (a) Loadings and (b) scores plots for fatty alcohols from the Ria Formosa lagoon after log transformation and PCA. Data from S.M. Mudge (unpublished).

on data which are defined as the signature such as the STP effluent seen in Figure 9.2.[162] The signature dataset, which can be based upon chemical concentrations, physical attributes or biological community information, is called the X-Block and ideally will be a pure source sample but could be made up of environmental samples which have a high proportion of a single source such as effluents from STPs or fatty alcohol product distributions from laboratory analysis. Since the samples come from the same source, although the concentrations may vary, PCA will generate a principal component 1 (PC1) that explains most of the variance in the analytical data, typically >90%. This projection or vector in n-dimensional space, where n is the number of chemical compounds analysed, can be described by a series of loading factors on each compound; those compounds which have a major impact on PC1 will have high loadings (either positive or negative) whereas those compounds which are relatively unimportant and, therefore, do not have a major influence on the data, will have values close to zero. PC2 is fitted orthogonal to the first component so there is no component of PC1 influencing PC2. Once the first two PCs have been elucidated, their projection can be described in terms of the two sets of loadings.

These projections, which represent the signature defined in terms of the chemical compounds used, can now be applied to the environmental data (Y-Block). The amount of variance explained or predicted by each X-Block signature can be quantified. This can be shown graphically either through a scatter plot of the weightings on each sample or as the total explained variance.[163] If the signature is similar to that of the environmental data, a high value for the explained variance is produced. Conversely, if a poor fit is produced, the explained variance is also small. Each signature can be fitted in turn and all are fitted independently of each other. If none of them explain the variation seen in the data, the fits will be small in every case. A fuller treatment of the PLS methodology including the matrix manipulations used can be found in Geladi and Kowalski.[162]

The advantage of PLS over other methods (*e.g.* simple ratios) is the way it uses all compounds and develops a signature based on the internal relationships between each one such as the data in Figure 9.7. In general, the more compounds that are used, the better the specificity of the signature. However, if several of the compounds are common between sources (*e.g.* all samples have lots of C_{16} which is nothing to do with the source discrimination), it may be better to reduce the number of compounds used to decrease the amount of overlap between signatures.[164]

PCA can be used independently of the PLS technique to determine the number of potential sources that may be present in the Y-Block. The scores plot from such an analysis will group sites according to their chemical composition; those that co-vary are likely to have the same or a similar source. Inspection of the groupings may provide an insight into the number of sources, although care must be exercised when dealing with mixtures of variable composition. This PCA technique may also be used to explore the source data and determine the groupings within the possible source materials.

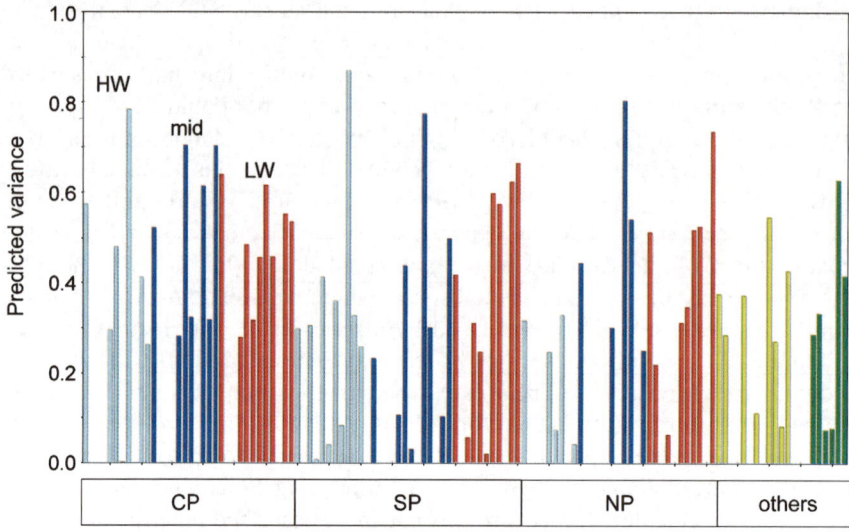

Figure 9.9   Amount of variance in beach samples (the Y-Block) using the fatty alcohols measured in faecal matter from domesticated animals (cows, *etc.*). CP, central pier; SP, south pier; NP, north pier. The pale blue bars are samples collected at the high water mark (HW), dark blue from the mid tide level (mid) and red from the low water mark (LW). Each bar is from a different week.

Using the fatty alcohol data obtained from cow faecal matter in the River Ribble catchment, which may contribute to Blackpool Beach, the amount of variance that that signature will predict in the beach samples (Y-Block) can be seen in Figure 9.9. These results show that for some samples the predictable variance is zero while for others it can be as high as 87%. Most results are around the 30% region. This implies that by using the fatty alcohols it is possible to predict this amount of variance in the beach samples by using the domesticated farm animals' faecal matter. This is due to the similarity between the fatty alcohol profile in the samples and the cow faeces. By looking at the PCA loadings plot (Figure 9.1), this is due to the amount of long chain compounds which load negatively on PC1.

Given a suitable set of samples which characterise the fatty alcohols derived from detergents, it would be possible to assess the contribution that this source made to any environmental fatty alcohol profile. As a first attempt, the data used to generate Figure 4.17 have been used to explain the profiles from the Ria Formosa lagoon in Portugal. The predictable variance can be seen in Figure 9.10 for each of the sample sites. The initial observation is that the values are considerably greater than might be expected in reality – to expect that more than 10% of the compounds come from detergents in such a system is not feasible. This highlights the inadequacy of using a simple chemical profile approach. As the chemicals used in the detergents are similar to those found in nature and in an area with high bacterial activity, a degree of overlap is to be

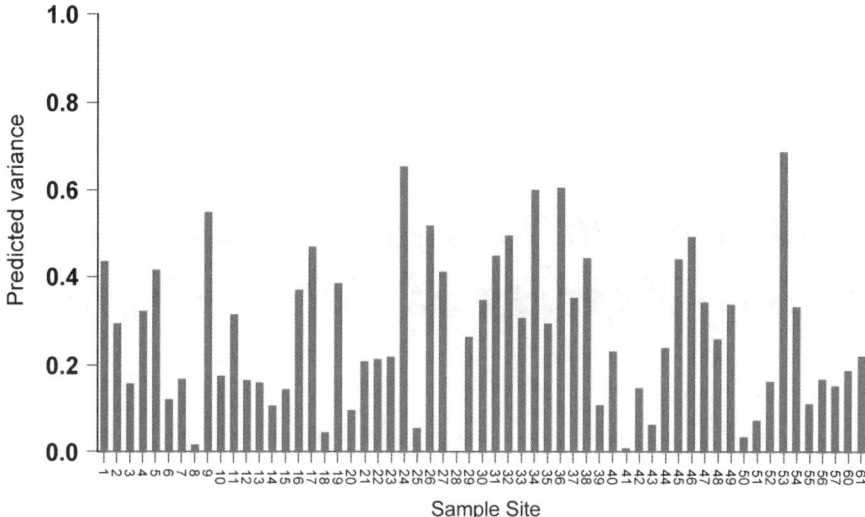

Figure 9.10   Amount of variance in a dataset from the Ria Formosa predicted from the alcohol signature from a series of European STPs. The values are significantly greater than might reasonably be expected.

expected. A better approach may be a constrained least squares approach such as that used by Burns *et al.*[165]

Further investigation of the use of signatures needs to be made so as to accurately predict the anthropogenic component in environmental samples. The best approaches for this may be through the use of compound-specific stable isotope analysis as there is likely to be significant differences between the marine produced compounds and those developed on land either as natural terrestrial materials or detergents.

Summary

- Simple biomarker approaches may not be sufficient to elucidate sources in complex environments with mixed inputs. In such situations, multivariate statistical methods may be useful.
- PCA can assist in identifying the clustering of compounds and sites and suggest the likely major organic matter source in samples.
- PLS can go one step further than PCA and quantify the source contributions, although in complex environments with similar source signatures even this may overestimate the contributions.

CHAPTER 10

Environmental and Human Safety Aspects of Fatty Alcohols

Are there potential toxic or ecotoxic effects from the use of fatty alcohols in consumer products?

10.1 Physical Chemistry Relevant to Safety Assessments

10.1.1 OECD SIAR Summary

Long chain fatty alcohols ($>C_6$ in this definition) were identified as a group of chemicals for the purpose of producing a complete human health and environmental safety dossier under the Organisation for Economic Cooperation and Development (OECD) High Production Volume (HPV) Chemicals Programme. The global chemical industry, through the International Council of Chemical Associations (ICCA), launched a voluntary initiative on HPV chemicals in 1998 to expedite the OECD HPV programme. Through this commitment, the chemical industry has undertaken to provide, as a first step, harmonised datasets on the intrinsic hazards and initial hazard assessments for approximately 1000 HPV substances. The information, consisting of a screening information dataset (SIDS) dossier, a SIDS initial assessment report (SIAR) and a SIDS initial assessment profile (SIAP), is submitted to the OECD for international agreement. A similar voluntary programme operated by the United States Environmental Protection Agency is making publicly available human health and environmental effects data from industry on an additional 1400 HPV chemicals.

In order to facilitate assessments with chemically related compounds, the OECD produced guidance on the establishment of "chemical categories" which is generically known as the "category approach". The long chain aliphatic alcohol category is based upon a homologous series of increasing

Table 10.1 Long chain alcohols (C_6–C_{22} primary and essentially linear) covered under the OECD long chain alcohols category submission.[166]

Chemical name	CAS no.	Chemical name	CAS no.
1-Hexanol	111-27-3	Alcohols, C_{16}–C_{18}	67762-27-0
1-Octanol	111-87-5	Alcohols, C_{14}–C_{18}	67762-30-5
1-Decanol	112-30-1	Alcohols, C_{10}–C_{16}	67762-41-8
1-Undecanol	112-42-5	Alcohols, C_8–C_{18}	68551-07-5
1-Tridecanol	112-70-9	Alcohols, C_{14}–C_{16}	68333-80-2
1-Tetradecanol	112-72-1	Alcohols, C_6–C_{12}	68603-15-6
1-Pentadecanol	629-76-5	Alcohols, C_{12}–C_{16}	68855-56-1
1-Hexadecanol	36653-82-4	Alcohols, C_{12}–C_{13}	75782-86-4
1-Eicosanol	629-96-9	Alcohols, C_{14}–C_{15}	75782-87-5
1-Docosanol	661-19-8	Alcohols, C_{12}–C_{14}	80206-82-2
Alcohols, C_{12}–C_{15}	63393-82-8	Alcohols, C_8–C_{10}	85566-12-7
Alcohols, C_9–C_{11}	66455-17-2	Alcohols, C_{10}–C_{12}	85665-26-5
Alcohols, C_{12}–C_{18}	67762-25-8	Alcohols, C_{18}–C_{22}	97552-91-5
9-Octadecen-1-ol (9Z)	143-28-2	Alcohols, C_{14}–C_{18} and C_{16}–C_{18} unsaturated	68155-00-0
Alcohols, C_{16}–C_{18} and C_{18} unsaturated	68002-94-8	Tridecanol, branched and linear	90583-91-8

carbon chain lengths sharing common and predictable physicochemical properties within the family covering a carbon (C) chain length range of C_6 to C_{22}.[5] A chemical category is formally defined as "a group of chemicals whose physicochemical and toxicological properties are likely to be similar or follow a regular pattern as a result of structural similarity".[166] The long chain alcohols (LCOH) category covered 30 CAS numbers (Table 10.1) and were evaluated as a cohesive group by OECD and its member countries in 2006.

This group covers aliphatic alcohols of commercial importance or relevance in a wide variety of industrial, commercial and institutional uses.[74] Many of the applications utilise the alcohol as an intermediate which is then sulfated, ethoxylated or propoxylated to make various types of detergent surfactants. Many applications utilise the free alcohol *per se* because of the emulsifying and lubricating properties of these materials. Modler *et al.*[74] cite that these uses include cosmetics, toiletries, pharmaceuticals and lubricant preparations. Chain lengths shorter and longer than the ones considered here and in this chapter on safety are outside the realm of the category due to increasing departure of the desired properties exploited for their commercial uses described above and are not considered further.

10.1.2 Physicochemical Overview From the OECD SIAR

In order to be considered as a chemical category, the physicochemical properties of the group require inspection. Fisk *et al.*[5] present a comprehensive

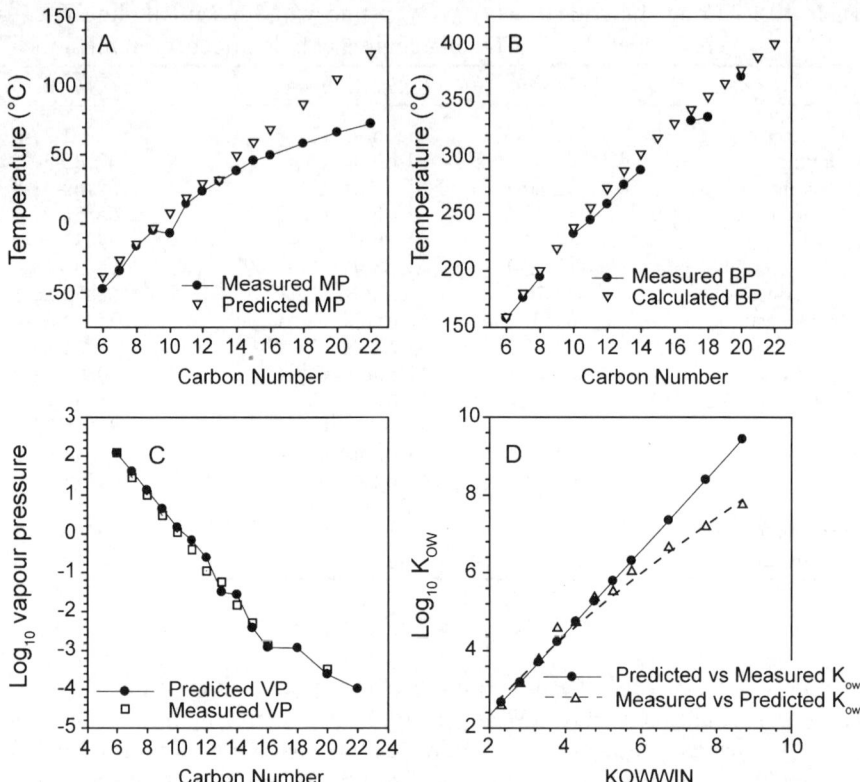

Figure 10.1   Relationships of measured and predicted physicochemical properties for long chain alcohols. MP, melting point; BP, boiling point; VP, vapour pressure. The K_{ow} values are adjusted for the carbon number. The relationship between the measured BP and predicted BP can be described by the equation $m_{BP} = 2.975 + 0.954\ p_{BP}$ with an R^2 of 0.996. Likewise, the measured VP can be predicted by $m_{VP} = -0.069 + 0.999\ p_{VP}$ with an R^2 of 0.973.

summary of toxicologically and environmentally relevant substance properties of long chain alcohols from C_6 to C_{22}. Wherever possible, measured values for the properties were assessed for reliability and were combined with modelled values to provide a comprehensive overview. The melting point, boiling point, vapour pressure and octanol–water partition coefficient were all found to be predictably ordered with respect to carbon number (Figure 10.1). For example, as carbon number increases, melting point increases. Departures from predictions were observed for some parameters, but were always directionally appropriate (*e.g.* K_{ow}). It is likely that this was reflecting additional co-variables beyond molecular weight or carbon number, such as the influence of solubility and sorptivity (Figure 10.2). Importantly, for the purpose of marrying physicochemical properties with emergent toxicological properties, a highly predictable relationship emerged (several of these are discussed below).

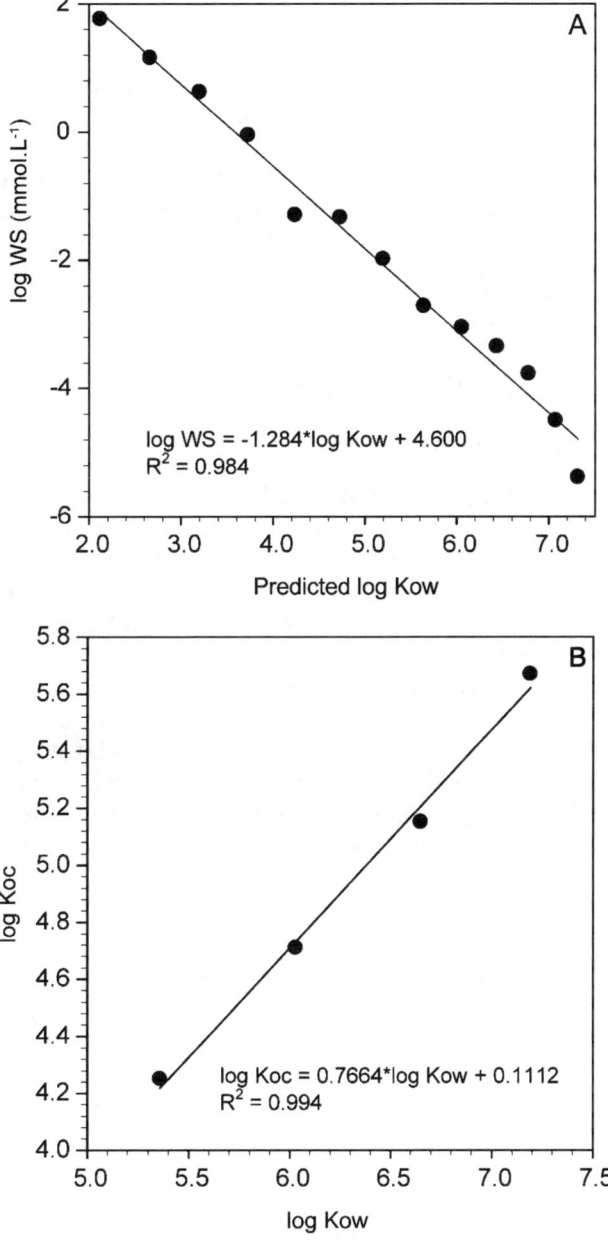

Figure 10.2 Relationship of the octanol–water partition coefficient (K_{ow}) with water solubility (WS) and partitioning to organic carbon (K_{oc}).

10.1.3 Conclusions for Characterising the OECD Long Chain Alcohols Category

A category approach is only justified when the physicochemical properties, which can be internally highly ordered, can be related to toxicological or environmental properties. In the case of the long chain alcohols category, this indeed was the case as determined at the OECD Screening Information Assessment Meeting (SIAM).[166] The remainder of this chapter will be devoted to discussing the known toxicology of the category and exposure to both humans and the environment.

10.2 Human Health and Risk

10.2.1 Uses and Products

Considerations for the assessment of long chain alcohols and human health are driven by the use in products and the potential routes of exposure. The category is relatively rich in empirical data as demonstrated in the OECD SIAR.[166] The category SIDS was presented as a whole set, rather than a substance-by-substance accounting, in order to facilitate broad justifiable conclusions for human health and exposure.

Consumer exposure scenarios were built for laundry detergents, fabric conditions and various personal care products in order to refine the assessments. In general, the skin is the dominant route of exposure associated with these products as would be projected from physicochemical properties such as vapour pressure. In order to be appropriately conservative, scenarios were also built for potential inhalation exposure through the use of sprays where long chain alcohols may be present.

The general exposure model takes the form:

Potential chemical exposure (PE) = Exposure to product (EXP)

$\times$ concentration in product formulation

This general equation is then modified to suit the various input needs for dermal or inhalation exposure. For example, it is possible to model or estimate exposure of the residual alcohol after skin application for a personal care product:

$$PE = \frac{[FQ \times A \times PR \times PT \times CF \times DA]}{BW} \times PF$$

or for a laundry product residual on clothing:

$$PE = \frac{[FQ \times CA \times PC \times FT \times CF \times TF \times DA]}{BW} \times PF$$

where FQ = frequency of use (use per day); CA = body surface contact area (cm^2); PC = product concentration $(g\,cm^{-3})$; FT = film thickness on skin (cm); TF = time scaling factor (unitless); DA = dermal absorption (%); BW = body weight (kg); PF = Fatty alcohol concentration in product formulation (%); A = amount per use (g per day or g per wash); PR = percent retained on clothing or on skin (%); PT = percent transferred from clothing to skin (%); CF = conversion factor (g → mg).

Similarly, it is possible to estimate exposure for inhalation of exposure to a spray or to undiluted product used as a hard surface cleaner.[167,168] The form and assumptions of these equations are based on first principles of safety and are mathematically consistent with published exposure guidelines used by regulators. Veenstra *et al.*[169] have provided a comprehensive overview of the wide array of consumer products considered regarding exposure to long chain fatty alcohols.

10.2.2 Hazards for Human Health

The mammalian metabolism of primary alcohols is a key factor in understanding the potential toxicology of individual components. The initial step in mammalian metabolism is oxidation to the corresponding carboxylic acid which is degraded further by acyl-CoA intermediates by the mitochondrial β-oxidation process (see Chapter 5). Two-carbon units are sequentially removed in a stepwise fashion. This evolutionarily conserved approach to alcohol metabolism occurs throughout the microbial world and in plants and animals (see also Chapters 2 and 3).

Veenstra *et al.*[169] state that "Aliphatic alcohols have a potential for absorption at varying degrees by all common routes of exposure. Based on comparative *in vitro* skin permeation data and dermal absorption studies in hairless mice, aliphatic alcohols show an inverse relationship between absorption potential and chain length with the shorter chain alcohols having a higher absorption potential than the longer chained alcohols [ref. 170 cited] consistent with the established relationship between skin penetration and physicochemical properties." However, quantitative data for dermal absorption have not been available in the literature and newly developed data demonstrated that the *in vitro* skin permeation of radiolabelled tetradecanol (C_{14}) can be modelled as a square-root-of-time function.[171] The dermal penetration study discussed by Veenstra *et al.* represents a "worst case" for dermal penetration of pure materials as chain lengths below tetradecanol (C_{14}) are unsuitable for the function they play in products and chain lengths above are sequentially less and less bioavailable due to molecular size and solubility.

Veenstra *et al.*[169] provide a serial description of the potential effects from exposure to long chain alcohols regarding acute toxicity (oral, inhalation and dermal), irritation and sensitisation (by inhalation, on skin and to eyes) and repeated dose toxicity (chronic exposure by inhalation, oral, intraperitoneal injection), mutagencitiy, carcinogenicity, reproductive toxicity and developmental toxicity. This large review encompassing hundreds of records demonstrates:

- Aliphatic alcohols are of a low order of toxicity upon single or repeated exposure.
- An inverse relationship exists between chain length and toxicity with shorter chain materials producing more pronounced effects.
- Long chain alcohols do not have skin sensitisation potential, are not mutagenic, carcinogenic, or developmental and reproductive toxins.
- The key human health hazards for the category are skin and eye irritation. For the aliphatic alcohols in the range C_6–C_{11} a potential for skin and eye irritation exists, without concerns for tissue destruction or irreversible changes. Aliphatic alcohols in the range C_{12}–C_{16} have a low degree of skin irritation potential; alcohols with chain lengths of C_{18} and above are non-irritants to skin. The eye irritation potential for alcohols with a chain length of C_{12} and above has been shown to be minimal.

10.2.3 Exposure Characterisation for Human Health

The general form of the exposure assessment has been described as a function of exposure to the product multiplied by the concentration of the chemical of interest in the product. A detailed survey on the use and exposure of alcohols was conducted by member companies of the Long Chain Aliphatic Alcohols Consortium under the leadership of The Soap and Detergent Association (U.S.).[172] Table 4.1 presents results of the survey which includes responses from North America, Europe and Asia Pacific geographic regions. These data formed the basis for exposure model inputs described by Veenstra *et al.*[169]

Exposure estimates, calculated for dermal penetration and inhalation routes of exposure, for more than a dozen categories of consumer products have been determined (Table 10.2). Dermal exposure estimates for laundry detergents and surface cleaners were all below $1 \mu g\,kg^{-1}$ body weight per day and inhalation exposure estimates were 2 orders of magnitude or more less than those of dermal exposure. Exposure estimates from personal care and cosmetic products are inherently larger due to the intended application of products to the body (Table 10.3).

10.2.4 Risk Characterisation for Human Health

Common practice in human health and safety assessment is to express the potential risk in terms of the margin of exposure (MoE). Typically this will consist of a ratio of the most relevant no adverse effect level (NOAEL) from toxicological studies for repeated exposure and screening level exposure estimates. Relatively large MoEs are sought which then obviate the need for additional uncertainty factors that can vary depending on the size, quantity and availability of the hazard data.[168,169]

Table 10.2 Estimates for dermal and inhalation exposure routes for various cleaning product categories.[169]

Exposure scenario	Exposure estimate[a] (mg kg^{-1} per day)	Margin of exposure
Dermal modeling		
Cleaning products		
Laundry pre-treatment (undiluted)	5.5×10^{-4}–2.8×10^{-3}	>10 000
Hand-wash of laundry (diluted)	4.3×10^{-5}–2.2×10^{-4}	>10 000
Hard surface cleaner (diluted)	1.3×10^{-4}–6.5×10^{-4}	>10 000
Hard surface cleaner (undiluted)	9.2×10^{-4}–4.6×10^{-3}	>10 000
Laundry product residual on clothing		
Liquid detergent	1.9×10^{-3}–9.2×10^{-3}	>10 000
Fabric conditioner	1.7×10^{-3}–8.6×10^{-3}	>10 000
Inhalation modeling		
Cleaning products		
Spray cleaner	3.6×10^{-6}–1.8×10^{-5}	>10 000

[a]Range based on minimum and maximum product concentration values.[172]

Table 10.3 Summary of the exposure estimates from the use of personal care and cosmetic products.[169]

Exposure scenario	Absorbed exposure estimate[a] (mg kg^{-1} per day)	Worst case margin of exposure[a]
Antiperspirants	0.5–1	1000
Body moisturiser	0.3–3	333
Cleansing products	8×10^{-3}–8×10^{-4}	125 000
Face/eye cosmetics	0.3–1	1000
Hair conditioner	10^{-2}–10^{-1}	10 000
Hair styling tonic/gel	7×10^{-2}–7×10^{-1}	1428

[a]Based on maximum formula concentration, maximum use amount, maximum skin penetration and conservative interpretation of NOAEL. Realistic exposures are anticipated to be lower and human equivalent NOAELs to be higher.

All MoEs calculated for laundry detergent and cleaning products are in excess of 10 000. Veenstra *et al.*[169] conclude that, taking into account the overall database for this category including the lack of concern regarding developmental and reproductive toxicity, genotoxicity and carcinogenicity for this category, the use of long chain alcohols in household products is without concern for human health.

For personal care products exposure, estimates were used that ensured the estimates presented worst case values.[169] True exposure is expected to be at least one order of magnitude less when actual product use scenarios and the types of alcohols used in each application are considered. An additional complication is that the primary biotransformation products of the aliphatic alcohols are indistinguishable from those derived from common dietary sources (triglycerides), with the human intake from dietary sources being several orders of magnitude above those arising from the use of long chain alcohols in personal care and cosmetic products.[169] As with cleaning products, the authors conclude that upon taking into account the overall database for this category, no concerns exist regarding developmental and reproductive toxicity, genotoxicity and carcinogenicity for this category justifying the conclusion that the use of long chain alcohols in personal care and cosmetic products is safe.

10.3   Environmental Risk

10.3.1   Pathways of Environmental Exposure

As discussed elsewhere in this book, both natural and anthropogenic sources of fatty alcohols are important in understanding environmental exposure from this group of chemicals. From the point of view of environmental risk, distinguishing natural sources from human-derived sources is relevant for assessing the need for risk management for the uses of alcohols.[166] The primary source of anthropogenic fatty alcohols to the environment is *via* disposal down drains of cleaning, laundry and personal care products to wastewater treatment systems following consumer use or unintended discharge at the point of manufacture. For drain disposal, validated models of sewage treatment plant operations can be used to characterise the potential presence in wastewater effluents and resulting exposures in the aquatic environment[173,174] which are then useful to characterise broad environmental exposure in riverine systems.[175,176] In the case of long chain alcohols which have a complex array of inputs to sewage and possess exceptional metabolic profiles, models become somewhat limited. Federle and Itrich[177] demonstrated exceptionally short half-lives of C_{12} and C_{16} alcohols in wastewater simulation tests. Mineralisation *via* biodegradation pathways to CO_2 and water dominated the losses and resulted in half-lives of the order of a few minutes. Figure 10.3 provides a schematic overview of the various sources and releases of long chain fatty alcohols from a sewage treatment point of view. A discussion of environmental monitoring of fatty alcohols for the purpose of environmental risk assessment follows in the context of known environmental toxicology of the category.

10.3.2   Environmental Effects

Fisk *et al.*[5] and Schäfers *et al.*[178] discuss acute and chronic toxicity of long chain alcohols to aquatic organisms. Fisk *et al.* demonstrate an elegant

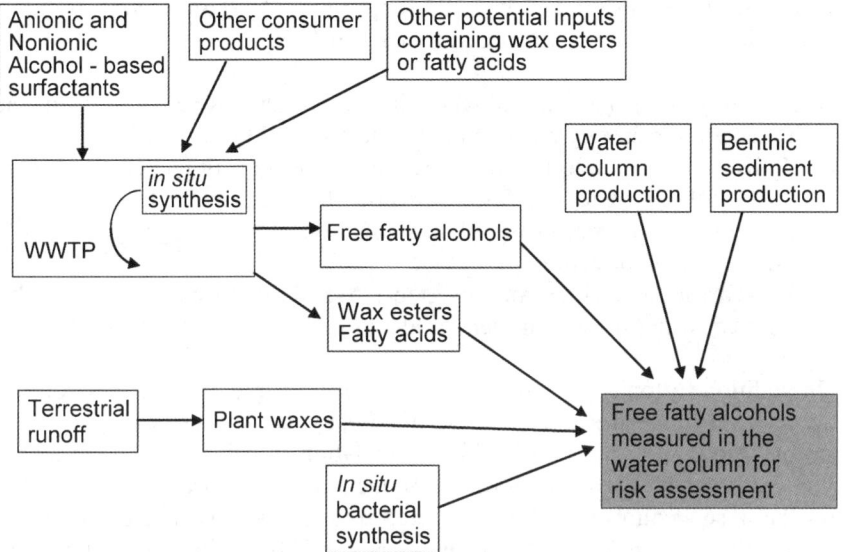

Figure 10.3 Schematic view of potential inputs and discharges from sewage treatment plants (WWTP) encompassing natural and anthropogenic sources of fatty alcohols. Adapted from Belanger *et al.*[179]

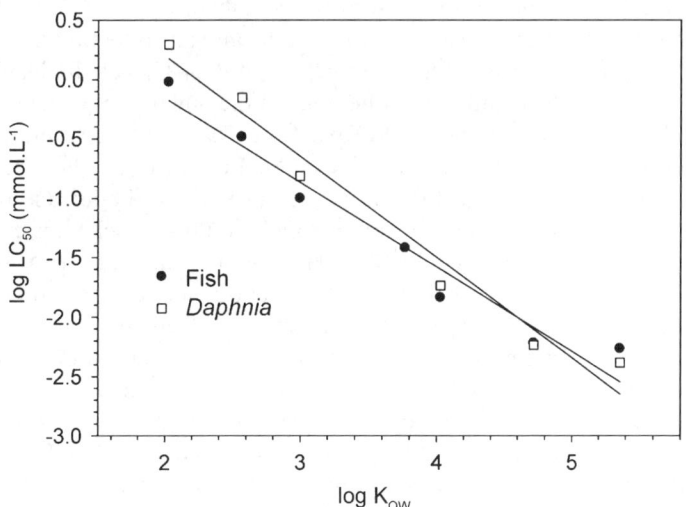

Figure 10.4 Acute structure–activity relationships for long chain alcohols for fish and *Daphnia*.[5,166]

structure–activity relationship (SAR) exists for both invertebrates (*Daphnia magna*, the water flea) and fish using a combination of quantitative SAR (QSAR) and measured, reliable toxicity values (Figure 10.4). Algae were not as data-rich as invertebrates and fish so conclusions, while qualitatively similar, were not suitable for summary in an SAR approach. The conclusions from Fisk *et al.* with regards to environmental toxicity include:

- Solubility limits the effects observed in many studies. Reliable acute toxicity information can only be generated from C_{12} and lower chain lengths.
- Hydrophobicity, reflected in the octanol–water partition coefficient (K_{ow}), is a useful parameter for describing potential toxicity.
- The QSARs for acute toxicity as a function of K_{ow} suggest a non-polar narcotic mode of action.
- Algae are least sensitive, and invertebrates and fish are equi-sensitive, when exposed to the same chain length alcohol.

In addition to toxicity relationships, Fisk *et al.*[5] explored other environmentally relevant properties as a function of the physical chemistry of the alcohol category. Similar to acute toxicity, bioaccumulation and biodegradability are also somewhat related to hydrophobicity, although here there are greater complications because alcohols are biotransformed by microbes and higher organisms. Overall, the biotransformation potential is a highly positive attribute of the alcohols because these processes limit environmental exposure and dispersion.

Schäfers *et al.*[178] addressed the challenges of developing chronic aquatic toxicity data for several long chain alcohols. Due to the high rate of degradation, maintenance of exposure levels in chronic toxicity tests is exceptionally difficult. In addition, the authors described the difficulties of working with very low solubility substances. A series of 21-day chronic reproduction and survival tests with *Daphnia magna* were conducted on C_{10}–C_{15} chain length fatty alcohols, many of which had to be conducted at the junction of water solubility. Table 10.4 summarises the measured toxicity to *Daphnia magna* and also suggests the difficulty in testing these substances. Free fatty alcohols are highly biodegradable and were subject to 100% loss in less than 24 hours (studies were conducted using daily renewals as flow-through designs, which were even more difficult to control). While Table 10.4 expresses the data as mean measured concentrations, it is also clear that initial concentrations were 3–6 times higher than means that included 24-hour-old exposures. Effects, measured as the 21-day EC_{10}, the concentration that effectively reduced reproduction by 10% relative to the control, were dose dependent and exposures were conducted right at the theoretical limit of solubility. An important observation was that C_{14} (tetradecanol) produced the lowest EC_{10} and that the EC_{10} for C_{15} (pentadecanol) was higher. This deflection in the SAR is expected when exposures pass the limit of solubility; the biological effects are no longer observed as the portion of the exposure that is soluble is not enough to incur toxicity.[180,181] Tests at chain lengths above C_{15} are not only physically impossible, but would yield no additional insight into effects of long chain fatty alcohols as these chain lengths are progressively more insoluble.

Schäfers *et al.*[178] concluded that even the C_{15} toxicity data were likely to be a mixture of both actual ecotoxicity and physical effects, although the authors attempted to minimise the latter. When the C_{15} data were not included, the

Table 10.4 Response of *Daphnia magna* to long chain fatty alcohols in 21 days, chronic reproduction and survival test.[178] Measured EC_{10} values were used when available and predictions for untested chain lengths (C_{13}) were based on (Q)SAR.

Chain length	Reproduction EC_{10} ($\mu g\,l^{-1}$)[a]	Water solubility[b] ($\mu g\,l^{-1}$)
Based on overall mean measured concentrations		
10	210 (1.33)	8000
12	12.8 (0.07)	1930
13	17.2 (0.09)	380
14	6.3 (0.03)	191
15	12 (0.05)	102
Based on mean initial measured concentrations		
10	610 (3.85)	8000
12	150 (0.81)	1930
13	148 (0.74)	380
14	70 (0.33)	191
15	74 (0.32)	102

[a]Values in parentheses in μmol.
[b]From Fisk *et al.*[5]

remaining studies suggested a QSAR slope that was consistent with a non-polar mode of action.

10.4 Measurements of Exposure for the Purpose of Environmental Risk Assessment

An alternative approach to generic modelling is to monitor concentrations in effluents under a variety of conditions. Several researchers[179,182,183] monitored long chain fatty alcohols (C_{12}–C_{18}) in wastewater treatment plant effluents. A wide array of treatment plant types (trickling filter, oxidation ditch, lagoon, activated sludge, rotating biological contactors) were assessed across Europe, Canada and the USA. The specific goal of these studies was to develop a frequency distribution of the presence of alcohols in 100% wastewater effluent to use in developing environmental risk assessment scenarios. The same sampling and analytical methods were used in all studies.[166] Belanger *et al.*[179] explain that the analytical methodology employed in these studies measures both "free" aliphatic alcohol (that which is not sorbed to effluent solids) and bound fatty alcohol and does not discriminate between natural and anthropogenic. Thus, the amount of measured fatty alcohol can be considered "worst case" with respect to environmental exposure to anthropogenic fatty alcohols. As a means to provide additional realism, Belanger *et al.*[179] also modelled the amount of alcohol that was bioavailable in solution using an alcohol-specific sorption (Q)SAR similar to that of Fisk *et al.*[5] In addition, the authors considered in-stream die-away kinetics from Federle and Itrich[177] and dilution in

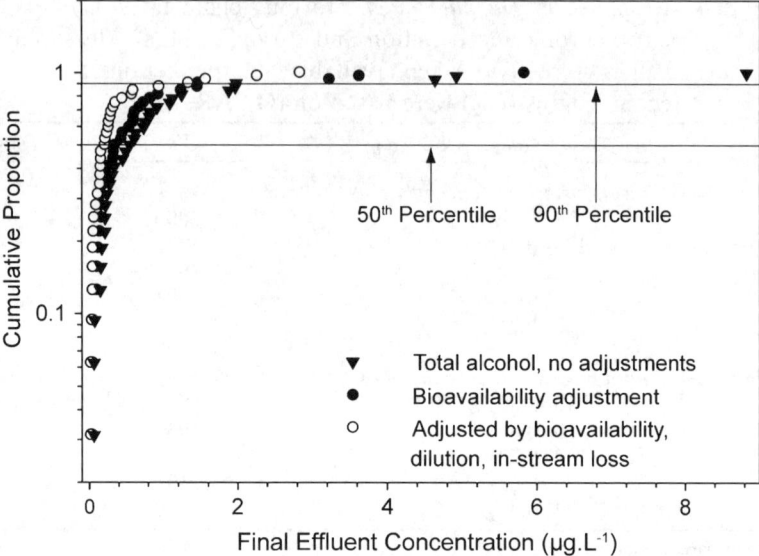

Figure 10.5   Cumulative frequency distribution of final effluent concentrations before (total, unadjusted) and after (total adjusted) correcting for bioavailability, dilution and in-stream mineralisation of parent alcohol.[179]

surface water. Figure 10.5 summarises the measured distribution for the 32 monitored sites across the globe.

10.5   Risk Characterisation in the Environment

As in human health and safety assessment, risk characterisation for the environment combines an evaluation of the predicted exposures and hazard, typically expressed as the PEC/PNEC ratio (the predicted environmental concentration divided by the predicted no-effect concentration in the environment). PEC/PNEC or risk characterisation ratios (RCRs) above 1.0 suggest a potential risk to the environment. It should be kept in mind that these represent *potential* risks, as testing and measurement methods are conducted in a way to maximise the ability to observe responses. Sensitive test species are used, exposures occur in clean water with no mitigating factors and application factors are applied to account for additional sources of uncertainty. Belanger *et al.*[179] synthesised monitoring studies with the work of Fisk *et al.*[5] and Schäfers *et al.*[178] Because alcohols are present in the environment as mixtures, Belanger *et al.* argue that it is most appropriate to take a mixture approach that sums the contributions for exposure and effects by chain length thereby producing a unified assessment for the entire category. A similar approach was performed for the ethoxylated alcohols.[182] Because chain lengths above C_{15}

are not expected to impart toxicity because of solubility–toxicity relationships developed in Schäfers *et al.*,[178] only chain lengths between C_{12} and C_{15} were considered for the environmental risk assessment. An application factor of 10 was applied to chronic *Daphnia* EC_{10} data to derive the PNEC and the PEC was based on 100% effluent exposure following adjustments for in-stream dilution and bioavailability. The highest PEC/PNEC (RCRs) of 0.09–0.59 were from sites employing fixed biofilm treatment technologies (such as trickling filters) whereas activated sludge treatment resulted in very small PEC/PNEC ratios of 0.004–0.17. Overall, Europe, the USA and Canada had RCRs of 0.03, 0.03 and 0.06, respectively, indicating a significant margin to conclude long chain alcohols are safe in the aquatic environment. Because the analytical methods cannot distinguish natural from anthropogenic sources and both are included in the environmental risk assessment through exposure monitoring, there is even greater confidence that anthropogenic sources pose very little risk.

Summary

- Fatty alcohols used in consumer products have a good human health profile and margins of exposure are most often in excess of 10 000. Fatty alcohols are not carcinogenic, mutagenic or reproductive/developmental toxins.
- The water solubility is a governing factor in the potential for toxic or ecotoxic effects, and while several of the detergent range alcohols fall into this category, effects beyond C_{15} are minimal.
- The environmental risk of fatty alcohols derived from consumer products such as detergents and cosmetic products is difficult to separate from fatty alcohols from the many natural sources. Sewage treatment plants essentially integrate fatty alcohols from all of these sources.
- The EC_{10} values for reproduction in *Daphnia magna* reached a minimum at C_{14}, near the limit of water solubility measurements. The predicted environmental concentrations divided by the predicted no-effect concentrations in the environment were all less than 1.0 and suggest that fatty alcohols do not pose a risk to the environment.

References

1. Y. Chikaraishi, K. Matsumoto, N. O. Ogawa, H. Suga, H. Kitazato and N. Ohkouchi, Hydrogen, carbon and nitrogen isotopic fractionations during chlorophyll biosynthesis in C3 higher plants, *Phytochemistry*, 2005, **66**(8), 911–920.
2. G. Kattner, C. Albers, M. Graeve and S. B. Schnack-Schiel, Fatty acid and alcohol composition of the small polar copepods, Oithona and Oncaea: indication on feeding modes, *Polar Biol.*, 2003, **26**(10), 666–671.
3. S. J. Ju and H. R. Harvey, Lipids as markers of nutritional condition and diet in the Antarctic krill *Euphausia superba* and *Euphausia crystallorophias* during austral winter, *Deep-Sea Res. II: Topical Stud. Oceanogr.*, 2004, **51**(17–19), 2199–2214.
4. J. J. Perry, J. T. Staley and S. Lory, *Microbial Life*, Sinauer Associates, Sunderland, MA, 2002.
5. P. Fisk, R. Wildey, A. Girling, H. Sanderson, S. Belanger, C. Schäfers, G. Veenstra, A. Nielsen, Y. Kasai, A. Willing and S. Dyer, Environmental properties of long-chain alcohols: 1. Physicochemical, environmental fate and acute aquatic toxicity properties, *Ecotoxicol. Environ. Safety*, 2008, in press.
6. L. P. Burkhard, D. W. Kuehl and G. D. Veith, Evaluation of reverse phase liquid-chromatography mass-spectrometry for estimation of n-octanol water partition-coefficients for organic-chemicals, *Chemosphere*, 1985, **14**(10), 1551–1560.
7. Y. B. Tewari, M. M. Miller, S. P. Wasik and D. E. Martire, Aqueous solubility and octanol water partition coefficient of organic compounds at 25.0 degrees C, *J. Chem. Eng. Data*, 1982, **27**(4), 451–454.
8. C. Hansch, J. E. Quinlan and G. L. Lawrence, Linear free-energy relationship between partition coefficients and the aqueous solubility of organic liquids, *J. Org. Chem.*, 1968, **33**(1), 347–350.
9. R. van Compernolle, D. McAvoy, A. Sherren, T. Wind, M. L. Cano, S. E. Belanger, P. B. Dorn and K. M. Kerr, Predicting the sorption of fatty alcohols and alcohol ethoxylates to effluent and receiving water solids, *Ecotoxicol. Environ. Safety*, 2006, **64**, 61–74.
10. T. W. Federle and N. R. Itrich, Fate of free and linear alcohol-ethoxylate-derived fatty alcohols in acvtivated sludge, *Ecotoxicol. Environ. Safety*, 2006, **64**, 30–41.

11. C. O. Rock and J. E. Cronan, Escherichia coli as a model for the regulation of dissociable (type II) fatty acid biosynthesis, *Biochim. Biophys. Acta: Lipids Lipid Metab.*, 1996, **1302**(1), 1–16.
12. A. L. Lehninger, D. L. Nelson and M. M. Cox, *Principles of Biochemistry*, Worth Publishers, New York, 1993.
13. T. Kaneda, Fatty acids of the genus, *Bacillus J. Bacteriol.*, 1967, **93**, 894–903.
14. P. E. Kolattukudy, R. Croteau and J. S. Buckner, Biochemistry of plant waxes, in *Chemistry and Biochemistry of Natural Waxes*, ed. P. E. Kolattukudy, Elsevier, Amsterdam, 1976, pp. 289–347.
15. M. Kates, Biosynthesis of lipids in microorganisms, *Annu. Rev. Microbiol.*, 1966, **20**, 13–44.
16. J. M. Berg, J. L. Tymoczko and L. Stryer, *Biochemistry*, W.H. Freeman & Co, New York, 2002.
17. J. G. Metz, M. R. Pollard, L. Anderson, T. R. Hayes and M. W. Lassner, Purification of a jojoba embryo fatty acyl-coenzyme A reductase and expression of its cDNA in high erucic acid rapeseed, *Plant Physiol.*, 2000, **122**(3), 635–644.
18. J. B. Cheng and D. W. Russell, Mammalian wax biosynthesis: I. Identification of two fatty acyl-coenzyme A reductases with different substrate specificities and tissue distributions, *J. Biol. Chem.*, 2004, **279**(36), 37789–37797.
19. J. B. Cheng and D. W. Russell, Mammalian wax biosynthesis: II. ERxpression cloning of wax synthase cDNAs encoding a member of the acyltransferase enzyme family, *J. Biol. Chem.*, 2004, **279**(36), 37798–37807.
20. J. E. Bishop and A. K. Hajra, Mechanism and specificity of formation of long-chain alcohols by developing rat brain, *J. Biol. Chem.*, 1981, **256**, 9542–9550.
21. A. A. Khan and P. E. Kolattukudy, microsomal fatty-acid synthetase coupled to acyl-CoA reductase in Euglena gracilis, *Arch. Biochem. Biophys.*, 1973, **158**(1), 411–420.
22. P. E. Kolattukudy and L. Rogers, Biosynthesis of fatty alcohols, alkane-1,2-diols and wax esters in particulate preparations from uropygial glands of white-crowned sparrows (Zonotrichia-Leucophrys), *Arch. Biochem. Biophys.*, 1978, **191**(1), 244–258.
23. P. E. Kolattukudy and L. Rogers, Acyl-CoA reductase and acyl-CoA fatty alcohol acyl transferase in the microsomal preparation from the bovine meibomian gland, *J. Lipid Res.*, **27**(4), 404–411.
24. X.-Y. Wu, R. A. Moreau and P. K. Stumpf, Studies of biosynthesis of waxes by developing jojoba seed: III. Biosynthesis of wax esters from acyl-CoA and long-chain alcohols, *Lipids*, 1981, **6**, 897–902.
25. P. E. Kolattukudy, Reduction of fatty acids to alcohols by cell-free preparations of Euglena gracilis, *Biochemistry*, 1970, **9**(5), 1095–1102.
26. S. Reiser and C. Somerville, Isolation of mutants of *Acinetobacter calcoaceticus* deficient in wax ester synthesis and complementation of one mutation with a gene encoding a fatty acyl coenzyme A reductase, *J. Bacteriol.*, 1997, **179**, 2969–2975.

27. J. Vioque and P. E. Kolattukudy, Resolution and purification of an aldehyde-generating and an alcohol-generating fatty acyl-CoA reductase from pea leaves (Pisum sativum L), *Arch. Biochem. Biophys.*, 1997, **340**(1), 64–72.

28. X. Wang and P. E. Kolattukudy, Solubilization and purification of aldehyde-generating fatty acyl-CoA reductase from green alga Botryococcus braunii, *FEBS Lett.*, 1995, **370**(1–2), 15–18.

29. G. Kattner and M. Graeve, Wax ester composition of the dominant calanoid copepods of the Greenland Sea Fram Strait region, *Polar Rese.*, 1991, **10**(2), 479–485.

30. G. Kattner and M. Krause, Seasonal variations of lipids (wax esters, fatty-acids and alcohols) in calanoid copepods from the North Sea, *Mar. Chem.*, 1989, **26**(3), 261–275.

31. J. R. Sargent and S. Falk-Petersen, The lipid biochemistry of calanoid copepods, *Hydrobiologia*, 1988, **167**, 101–114.

32. J. R. Sargent, R. F. Lee and J. C. Nevenzel, *Marine waxes*, in *Chemistry and Biochemistry of Natural Waxes*, ed. P. E. Kolattukudy, Elsevier, Amsterdam, 1976, pp. 49–91.

33. R. G. Ackman, C. S. Tocher and J. McLachlan, Marine phytoplankter fatty acids, *J. Fish. Res. Board Can.*, 1968, **25**, 1603–1620.

34. K. Kates and B. E. Volcani, Lipid components of diatoms, *Biochem. Biophys. Acta*, 1966, **116**, 264–278.

35. G. Kattner, G. Gercken and K. Eberlein, Development of lipids during a spring plankton bloom in the northern North Sea: I. Particulate fatty acids, *Mar. Chem.*, 1983, **14**, 149–162.

36. A. P. Tulloch, Chemistry of waxes of higher plants, in *Chemistry and Biochemistry of Natural Waxes*, ed. P. E. Kolattukudy, Elsevier, Amsterdam, 1976, pp. 235–287.

37. K. A. Dahl, D. W. Oppo, T. I. Eglinton, K. A. Hughen, W. B. Curry and F. Sirocko, Terrigenous plant wax inputs to the Arabian Sea: implications for the reconstruction of winds associated with the Indian monsoon, *Geochim. Cosmochim. Acta*, 2005, **69**(10), 2547–2558.

38. J. S. Buckner, M. C. Mardaus and D. R. Nelson, Cuticular lipid composition of Heliothis virescens and Helicoverpa zea pupae, *Comp. Biochem. Physiol., B: Biochem. Mol. Biol.*, 1996, **114**(2), 207–216.

39. D. R. Nelson, C. L. Fatland, J. S. Buckner and T. P. Freeman, External lipids of adults of the giant whitefly, Aleurodicus dugesii, *Comp. Biochem. Physiol., B: Biochem. Mol. Biol.*, 1999, **123**(2), 137–145.

40. J. Jacob, Bird waxes, in *Chemistry and Biochemistry of Natural Waxes*, ed. P. E. Kolattukudy, Elsevier, Amsterdam, 1976, pp. 93–146.

41. Y. M. Zhang, Y. J. Lu and C. O. Rock, The reductase steps of the type II fatty acid synthase as antimicrobial targets, *Lipids*, 2004, **39**(11), 1055–1060.

42. P. W. Albro, Bacterial waxes, in *Chemistry and Biochemistry of Natural Waxes*, ed. P. E. Kolattukudy, Elsevier, Amsterdam, 1976, pp. 419–445.

43. R. B. Johns, G. J. Perry and K. S. Jackson, Contribution of bacterial lipids to recent marine sediments, *Estuar. Coast. Marine Sci.*, 1977, **5**(4), 521–529.

44. M. Kates, Bacterial lipids, *Adv. Lipid Res.*, 1964, **2**, 17–90.

45. W. M. O'Leary, The fatty acids of bacteria, *Bacteriol. Rev.*, 1962, **26**, 421–447.

46. T. G. Tornabene, E. Gelpi and J. Oro, Identification of the fatty acids and aliphatic hydrocarbons in Sarcina lutea by gas chromatography and combined gas chromatography–mass spectrometry, *J. Bacteriol.*, 1967, **94**, 333–343.

47. M. Waltermann, A. Hinz, H. Robenek, D. Troyer, R. Reichelt, U. Malkus, H. J. Galla, R. Kalscheuer, T. Stoveken, P. von Landenberg and A. Steinbuchel, Mechanism of lipid-body formation in prokaryotes: how bacteria fatten up, *Mol. Microbiol.*, 2005, **55**(3), 750–763.

48. R. Kalscheuer and A. Steinbuchel, A novel bifunctional wax ester synthase/acyl-CoA:diacylglycerol acyltransferase mediates wax ester and triacylglycerol biosynthesis in Acinetobacter calcoaceticus ADP1, *J. Biol. Chem.*, 2003, **278**(10), 8075–8082.

49. J. Daniel, C. Deb, V. S. Dubey, T. D. Sirakova, B. Abomoelak, H. R. Morbidoni and P. E. Kolattukudy, Induction of a novel class of diacylglycerol acyltransferases and triacylglycerol accumulation in *Mycobacterium tuberculosis* as it goes into a dormancy-like state in culture, *J. Bacteriol.*, 2004, **186**, 5017–5030.

50. R. J. Parkes and J. Taylor, The relationship between fatty acid distributions and bacterial respiratory types in contemporary marine sediments, *Estuar. Coast. Shelf Sci.*, 1983, **16**(2), 173–174.

51. J. J. Boon, J. W. De Leeuw, G. J. V. d. Hoek and J. H. Vosjan, Significance and taxonomic value of iso and anteiso monoenoic fatty acids and branched b-hydroxy acids in Desulphovibrio desulfuricans, *J. Bacteriol.*, 1977, **129**, 1183–1191.

52. J. J. Boon, J. W. d. Leeuw and A. L. Burlingame, Organic geochemistry of Walvis Bay diatomaceous ooze: III. Structural analysis of the monoenoic and polycyclic fatty acids, *Geochim. Cosmochim. Acta*, 1978, **42**(6 Part 1), 631–644.

53. R. G. Leo and P. L. Parker, Branched chain fatty acids in sediments, *Science*, 1966, **152**, 649–650.

54. G. J. Perry, J. K. Volkman, R. B. Johns and J. Bavor, Fatty acids of bacterial origin in contemporary marine sediments, *Geochim. Cosmochim. Acta*, 1979, **43**(11), 1715–1725.

55. W. L. Jeng and C. A. Huh, Lipids in suspended matter and sediments from the East China Sea shelf, *Org. Geochem.*, 2004, **35**(5), 647–660.

56. S. M. Mudge and C. E. Norris, Lipid biomarkers in the Conwy Estuary (North Wales UK): a comparison between fatty alcohols and sterols, *Mar. Chem.*, 1997, **57**(1–2), 61–84.

57. E. W. Baker and J. W. Louda, Thermal aspects in chlorophyll geochemistry, in *Advances in Organic Geochemistry*, ed. M. Bjorøy, Wiley, Chichester, 1983, pp. 401–421.

58. F. R. Shuman and C. J. Lorenzen, Quantitative degradation of chlorophyll by a marine herbivore, *Limnol. Oceanogr.*, 1975, **20**, 580–586.

59. S. W. Jeffrey, Profiles of photosynthetic pigments in the ocean using thin-layer chromatography, *Marine Biology (Historical Archive)*, 1974, **26**(2), 101–110.

60. P. Cuny and J. F. Rontani, On the widespread occurrence of 3-methylidene-7,11,15-trimethylhexadecan-1,2-diol in the marine environment: a specific isoprenoid marker of chlorophyll photodegradation, *Mar. Chem.*, 1999, **65**(3–4), 155–165.

61. J. K. Volkman, S. M. Barrett, S. I. Blackburn, M. P. Mansour, E. L. Sikes and F. Gelin, Microalgal biomarkers: a review of recent research developments, *Org. Geochem.*, 1998, **29**(5–7), 1163–1179.

62. J.-P. Berge, J.-P. Gouygou, J.-P. Dubacq and P. Durand, Reassessment of lipid composition of the diatom, *Skeletonema costatum, Phytochemistry*, 1995, **39**(5), 1017–1021.

63. F. A. Abreu-Grobois, T. C. Billyard and T. J. Walton, Biosynthesis of heterocyst glycolipids of Anabaena cylindrical, *Phytochemistry*, 1977, **16**(3), 351–354.

64. T. Rezanka and M. Podojil, Identification of wax esters of the fresh-water green alga *Chlorella kessleri* by gas chromatography–mass spectrometry, *J. Chromatogr. A*, 1986, **362**, 399–406.

65. T. Rezanka, O. Vyhnalek and M. Podojil, Identification of sterols and alcohols produced by green algae of the genera *Chlorella* and *Scenedesmus* by means of gas chromatography–mass spectrometry, *Folia Microbiol.*, 1986, **31**, 44–49.

66. J. K. Volkman, S. M. Barrett, G. A. Dunstan and S. W. Jeffrey, C30–C32 alkyl diols and unsaturated alcohols in microalgae of the class Eustigmatophyceae, *Org. Geochem.*, 1992, **18**(1), 131–138.

67. A. Hayeememon, M. Shameel, M. Ahmad, V. U. Ahmad and K. Usmanghani, Phycochemical studies on *Gracilaria foliifera* (Gigartinales, Rhodophyta), *Botanica Marina*, 1991, **34**(2), 107–111.

68. H. A. M. Ali, R. W. Mayes, B. L. Hector and E. R. Orskov, Assessment of n-alkanes, long-chain fatty alcohols and long-chain fatty acids as diet composition markers: the concentrations of these compounds in rangeland species from Sudan, *Animal Feed Sci. Technol.*, 2005, **121**(3–4), 257–271.

69. C. J. Nott, Biomarkers in ombrotrophic mires as palaeoclimate indicators, *Chemistry*, Bristol University, 2000, p. 231.

70. S. C. Xie, C. J. Nott, L. A. Avsejs, D. Maddy, F. M. Chambers and R. P. Evershed, Molecular and isotopic stratigraphy in an ombrotrophic mire for paleoclimate reconstruction, *Geochim. Cosmochim. Acta*, 2004, **68**(13), 2849–2862.

71. C. E. Hamm and V. Rousseau, Composition, assimilation and degradation of *Phaeocystis globosa*-derived fatty acids in the North Sea, *J. Sea Res.*, 2003, **50**(4), 271–283.

72. J. K. Volkman, R. R. Gatten and J. R. Sargent, Composition and origin of milky water in the North Sea, *J. Mar. Biol. Assoc. UK*, 1980, **60**(3), 759–768.
73. K. L. Matheson, Surfactants raw materials: classification, synthesis, and uses, in *Soaps and Detergents: A Theoretical and Practical Review*, ed. L. Spitz, AOCS Press, Champaign IL, 1996, pp. 288–303.
74. R. F. Modler, R. Gubler and Y. Inoguchi, Detergent alcohols, *CEH Marketing Research Report*, 2004, pp. 1–71.
75. S. Deng, Sorbent technology, in *Encyclopedia of Chemical Processing*, ed. S. Lee, CRC Press, 2006, pp. 2825–2845.
76. A. E. Comyns, *Encyclopedic Dictionary of Named Processes in Chemical Technology*, CRC Press, 1999.
77. B. Vora, A. Bozzano and S. Sohn, Detergent alkylate, in *Encyclopedia of Chemical Processing*, ed. S. Lee, CRC Press, 2006, pp. 663–672.
78. J. Falbe, *New Syntheses With Carbon Monoxide*, Springer-Verlag, 1980.
79. M. Grant-Huyser, S. Maharaj, L. Matheson, L. Rowe and E. Sones, Ethoxylation of detergent-range OXO alcohols derived from Fischer–Tropsch α-olefins, *J. Surfactants Detergents*, 2004, **7**(4), 397–407.
80. W. W. Schmidt, D. M. Singleton and K. H. Raney, Solution and performance properties of new biodegradable high-solubilty surfactants, *Proc. CESIO*, 2000, **2**, 1085–1093.
81. R. F. Modler, Detergent alcohols, *CEH Marketing Research Report*, SRI Consulting, 2004, p. 16.
82. S. E. Belanger and P. B. Dorn, Chronic aquatic toxicity of alcohol ethoxylate (AE) surfactants under Canadian exposure conditions, 31st Annual Aquatic Toxicity Workshop, Charlottetown, Prince Edward Island, Canadian Technical Report of Fisheries and Aquatic Sciences, 2004.
83. C. V. Eadsforth, A. J. Sherren, M. A. Selby, R. Toy, W. S. Eckhoff, D. C. McAvoy and E. Matthijs, Monitoring of environmental fingerprints of alcohol ethoxylates in Europe and Canada, *Ecotoxicol. Environ. Safety*, 2006, **64**(1), 14–29.
84. S. W. Morrall, J. C. Dunphy, M. L. Cano, A. Evans, D. C. McAvoy, B. P. Price and W. S. Eckhoff, Removal and environmental exposure of alcohol ethoxylates in US sewage treatment, *Ecotoxicol. Environ. Safety*, 2006, **64**(1), 3–13.
85. T. Wind, R. J. Stephenson, C. V. Eadsforth, A. Sherren and R. Toy, Determination of the fate of alcohol ethoxylate homologues in a laboratory continuous activated-sludge unit study, *Ecotoxicol. Environ. Safety*, 2006, **64**(1), 42–60.
86. R. F. Modler, M. Blagoev and Y. Inoguchi, Detergent alcohols, *CEH Marketing Research Report*, 2007, p. 81.
87. K. L. Kaddam, Power plant flue gas as a source of CO_2 for microalgae cultivation: economic impact of different process options, *Energy Conserv. Manag.*, 1997, **38**(Suppl), S505–S510.
88. K. S. Dhugga, Maize biomass yield and composition for biofuels, *Crop Sci.*, 2007, **47**(6), 2211–2227.

89. J. Fargione, J. Hill, D. Tilman, S. Polasky and P. Hawthorne, Land clearing and the biofuel carbon debt, *Science*, 2008, **319**(5867), 1235–1238.
90. T. Searchinger, R. Heimlich, R. A. Houghton, F. Dong, A. Elobeid, J. Fabiosa, S. Tokgoz, D. Hayes and T.-H. Yu, Use of US croplands for biofuels increases greenhouse gases through emissions from land-use change, *Science*, 2008, **319**(5867), 1238–1240.
91. Y. Basiron, Palm oil production through sustainable plantations, *Eur. J. Lipid Sci. Technol.*, 2007, **109**(4), 289–295.
92. T. Kurevija and N. Kukulj, Global environmental issues concerning large scale biodiesel production, in *Energy and the Environment 2006, vol. 2, 20th International Congress on Energy and the Environment*, 2006, 197–206.
93. D. Fairless, Biofuel: the little shrub that could – maybe, *Nature*, 2007, **449**, 652–655.
94. A. Pickering, The oil reserves production relationship, *Energy Econ.*, 2008, **30**, 352–370.
95. Y. Chisti, Biodiesel from micro-algae, *Biotechnol. Adv.*, 2007, **25**, 294–306.
96. A. M. Henstra, J. Sipma, A. Rinzema and A. J. M. Stams, Microbiology of synthesis gas fermentation for biofuel production, *Curr. Opin. Biotechnol.*, 2007, **18**(3), 200–206.
97. J. Sheehan, T. Dunahay, J. Benemann and P. Roessler, A look back at the US Department of Energy's Aquatic Species Program: biodiesel from algae, 1998, p. 296.
98. P. Spolaore, C. Joannis-Cassan, E. Duran and A. Isambert, Commercial applications of microalgae, *J. Biosci. Eng.*, 2006, **101**, 87–96.
99. R. Kalscheuer, T. Stolting and A. Steinbuchel, Microdiesel: Escherichia coli engineered for fuel production, *Microbiology-Sgm*, 2006, **152**, 2529–2536.
100. C. Lee, S. Wakeham and C. Arnosti, Particulate organic matter in the sea: the composition conundrum, *Ambio*, 2004, **33**(8), 565–575.
101. M. Soltani, P. Metzger and C. Largeau, Effects of hydrocarbon structure on fatty acid, fatty alcohol, and beta-hydroxy acid composition in the hydrocarbon-degrading bacterium Marinobacter hydrocarbonoclasticus, *Lipids*, 2004, **39**(5), 491–505.
102. T. Ishige, A. Tani, Y. Sakai and N. Kato, Wax ester production by bacteria, *Curr. Opin. Microbiol.*, 2003, **6**(3), 244–250.
103. M. M. Roper, The isolation and characterisation of bacteria with the potential to degrade waxes that cause water repellency in sandy soils, *Aust. J. Soil Res.*, 2004, **42**(4), 427–434.
104. S. Caradec, V. Grossi, F. Gilbert, C. Guigue and M. Goutx, Influence of various redox conditions on the degradation of microalgal triacylglycerols and fatty acids in marine sediments, *Org. Geochem.*, 2004, **35**(3), 277–287.
105. L. L. D. Cooper, J. E. Oliver, E. D. De Vilbiss and R. P. Doss, Lipid composition of the extracellular matrix of *Botrytis cinerea* germlings, *Phytochemistry*, 2000, **53**(2), 293–298.
106. R. P. Doss, Composition and enzymatic activity of the extracellular matrix secreted by germlings of Botrytis cinerea, *Appl. Environ. Microbiol.*, 1999, **65**(2), 404–408.

107. K. L. Larsen, M. Miller and R. P. Cox, Incorporation of exogenons long-chain alcohols into bacteriochlorophyll-c homologs by Chloroflexus aurantiacus, *Arch. Microbiol.*, 1995, **163**(2), 119–123.

108. R. I. Haddad, C. S. Martens and J. W. Farrington, Quantifying early diagenesis of fatty-acids in a rapidly accumulating coastal marine sediment, *Org. Geochem.*, 1992, **19**(1–3), 205–216.

109. A. Hotham, Core profiles of biomarkers for land use change in salt and freshwater Scottish lochs, *Ocean Sciences*, University of Wales, Bangor, 2001, p. 56.

110. M. Mohd Ali, Multivariate statistical analyses in lipid biomarker studies, *Ocean Sciences,* University of Wales, Bangor, 2003, p. 277.

111. R. K. F. Unsworth, Sedimentary lipid and PAH biomarkers as temporal indicators of change within the western area of the Ria Formosa Lagoon, Portugal, *Ocean Sciences*, University of Wales, Bangor, 2001, p. 88.

112. S. M Mudge, M. Bebianno, J. A. East and L. A. Barreira, Sterols in the Ria Formosa lagoon, *Portugal, Water Res.*, 1999, **33**(4), 1038–1048.

113. S. M. Mudge, J. A. East, M. J. Bebianno and L. A. Barreira, Fatty acids in the Ria Formosa Lagoon, *Portugal, Org. Geochem.*, 1998, **29**(4), 963–977.

114. S. G. Wakeham, C. Lee, J. I. Hedges, P. J. Hernes and M. L. Peterson, Molecular indicators of diagenetic status in marine organic matter, *Geochim. Cosmochim. Acta*, 1997, **61**(24), 5363–5369.

115. R. B. Gagosian and E. T. Peltzer, The importance of atmospheric input of terrestrial organic matter to deep-sea sediments, *Org. Geochem.*, 1986, **10**, 661–669.

116. F. G. Prahl, L. A. Muelhausen and M. Lyle, An organic geochemical assessment of oceanographic conditions at MANOP Site C over the past 26 000 years, *Paleoceanography*, 1989, **4**, 495–510.

117. W. L. Jeng, C. A. Huh and C. L. Chen, Alkanol and sterol degradation in a sediment core from the continental slope of southwestern Taiwan, *Chemosphere*, 1997, **35**(11), 2515–2523.

118. M. Y. Sun and S. G. Wakeham, Molecular evidence for degradation and preservation of organic matter in the anoxic Black Sea Basin, *Geochim. Cosmochim. Acta*, 1994, **58**(16), 3395–3406.

119. M. Y. Sun, S. G. Wakeham and C. Lee, Rates and mechanisms of fatty acid degradation in oxic and anoxic coastal marine sediments of Long Island Sound New York, USA, *Geochim. Cosmochim. Acta*, 1997, **61**, 341–355.

120. S. M. Mudge, *The Source of Organic Matter on Blackpool Beaches (2000)*, University of Wales, Bangor, 2001.

121. N. R. Itrich and T. W. Federle, Effect of ethoxylate number and alkyl chain length on the pathway and kinetics of linear alcohol ethoxylate biodegradation in activated sludge, *Environ. Toxicol. Chem.*, 2004, **23**(12), 2790–2798.

122. N. S. Battersby, A. J. Sherren, R. N. Bumpus, R. Eagle and I. K. Molade, The fate of linear alcohol ethoxylates during activated sludge treatment, *Chemosphere*, 2001, **45**, 109–121.

123. L. Kravetz, H. Chung, K. F. Guin, W. T. Shebs and L. S. Smith, Primary and ultimate biodegradation of an alcohol ethoxylate and nonylphenol ethoxylate under average winter conditions in the Unites States, *Tenside Surfactants Detergents*, 1984, **21**, 1–6.

124. J. Steber and P. Wierich, The environmental fate of detergent range fatty alcohol ethoxylates: biodegradation studies with a ^{14}C labeled model surfactant, *Tenside Surfactants Detergents*, 1983, **20**, 183–187.

125. J. C. Dunphy, D. G. Pessler and S. W. Morrall, Derivatization LC/MS for the simultaneous determination of fatty alcohol and alcohol ethoxylate surfactants in water and wastewater samples, *Environ. Sci. Technol.*, 2001, **35**(6), 1223–1230.

126. S. M. Mudge and C. E. Duce, Identifying the source, transport path and sinks of sewage derived organic matter, *Environ. Pollut.*, 2005, **136**(2), 209–220.

127. J. Folch, M. Lees and G. H. S. Stanley, A simple method for the isolation and purification of total lipides from animal tissues, *J. Biol. Chem.*, 1957, **226**(1), 497–509.

128. Y. Chikaraishi and H. Naraoka, δ^{13}C and δD identification of sources of lipid biomarkers in sediments of Lake Haruna (Japan), *Geochim. Cosmochim. Acta*, 2005, **69**(13), 3285–3297.

129. S. J. Rowland, W. G. Allard, S. T. Belta, G. Massae, J. M. Robert, S. Blackburn, D. Frampton, A. T. Revill and J. K. Volkman, Factors influencing the distributions of polyunsaturated terpenoids in the diatom, Rhizosolenia setigera, *Phytochemistry*, 2001, **58**(5), 717–728.

130. S. J. Rowland, S. T. Belt, E. J. Wraige, G. Masse, C. Roussakis and J. M. Robert, Effects of temperature on polyunsaturation in cytostatic lipids of Haslea ostrearia, *Phytochemistry*, 2001, **56**(6), 597–602.

131. S. M. Mudge and D. G. Lintern, Comparison of sterol biomarkers for sewage with other measures in Victoria Harbour, BC, Canada, *Estuar. Coast. Shelf Sci.*, 1999, **48**(1), 27–38.

132. C. G. Seguel, S. M. Mudge, C. Salgado and M. Toledo, Tracing sewage in the marine environment: altered signatures in Conception Bay, Chile, *Water Res.*, 2001, **35**(17), 4166–4174.

133. S. M. Mudge and C. G. Seguel, Organic contamination of San Vicente Bay, Chile, *Mar. Pollut. Bull.*, 1999, **38**(11), 1011–1021.

134. L. A. D. S. Madureira, Lipids in recent sediments of the eastern North Atlantic, *Chemistry*, Bristol University, 1994, p. 246.

135. J. McEvoy, The origin and diagenesis of organic lipids in sediments from the San Miguel Gap, *Chemistry*, Bristol University, 1983, p. 507.

136. V. J. Howell, Organic geochemistry of sediments from legs 67, 71 and 72 of the Deep Sea Drilling project, *Chemistry*, University of Bristol, 1984, p. 159.

137. D. Nash, R. Leeming, L. Clemow, M. Hannah, D. Halliwell and D. Allen, Quantitative determination of sterols and other alcohols in overland flow from grazing land and possible source materials, *Water Res.*, 2005, **39**(13), 2964–2978.

138. A. Otto, C. Shunthirasingham and M. J. Simpson, A comparison of plant and microbial biomarkers in grassland soils from the Prairie Ecozone of Canada, *Org. Geochem.*, 2005, **36**(3), 425–448.

139. D. A. Pickering, Chemical and physical analysis of laminated sediment formed in Looe Pool, Cornwall, *Environmental Sciences*, Plymouth Polytechnic, 1987, p. 376.

140. J. A. Scott, Mountain lake sedimentary biomarker records as indicators of holocene climate variability, *Chemistry*, University of Bristol, 2004, p. 218.

141. S. M. Mudge and C. G. Seguel, Trace organic contaminants and lipid biomarkers in Concepcion and San Vicente Bays, *Boletin De La Sociedad Chilena De Quimica*, 1997, **42**(1), 5–15.

142. L. Farías, L. A. Chuecas and M. A. Salamanca, Effect of coastal upwelling on nitrogen regeneration from sediments and ammonium supply to the water column in Concepción Bay, Chile, *Estuar. Coast. Shelf Sci.*, 1996, **43**, 137–155.

143. A. Lepez, El emisario submarino como sistema de tratamiento de aguas servidas [Subtidal emission with a treatment system for service waters], ESSBIO SA, 1996, p. 19.

144. L. M. Barreira, M. J. Bebianno, S. M. Mudge, A. M. Ferreira, C. I. Albino and L. M. Veriato, Relationship between PCBs in suspended and settled sediments from a coastal lagoon, *Ciencias Marinas*, 2005, **31**(1B), 179–195.

145. A. Newton and S. M. Mudge, Lagoon–sea exchanges, nutrient dynamics and water quality management of the Ria Formosa (Portugal), *Estuar. Coast. Shelf Sci.*, 2005, **62**(3), 405–414.

146. S. M. Mudge, J. D. Icely and A. Newton, Oxygen depletion in relation to water residence times, *J. Environ. Monit.*, 2007, **9**(11), 1194–1198.

147. M. J. Bebianno, Effects of pollutants in the Ria-Formosa-Lagoon, Portugal, *Sci. Total Environ.*, 1995, **171**(1–3), 107–115.

148. D. M. John, G. E. Douglas, S. J. Brooks, G. C. Jones, J. Ellaway and S. Rundle, Blooms of the water net Hydrodictyon reticulatum (Chlorococcales, Chlorophyta) in a coastal lake in the British Isles: their cause, seasonality and impact, *Biologia*, 1998, **53**(4), 537–545.

149. J. E. Flory and G. R. W. Hawley, A *Hydrodictyon reticulatum* bloom at Looe Pool, *Cornwall, Eur. J. Phycology*, 1994, **29**(1), 17–20.

150. L. A. Avsejs, The organic geochemistry and compound specific radiocarbon dating of peat and other sedimentary materials. *Chemistry*, Bristol University, 2001, p. 211.

151. L. A. Avsejs, C. J. Nott, S. C. Xie, D. Maddy, F. M. Chambers and R. P. Evershed, 5-n-Alkylresorcinols as biomarkers of sedges in an ombrotrophic peat section, *Org. Geochem.*, 2002, **33**(7), 861–867.

152. C. J. Nott, S. C. Xie, L. A. Avsejs, D. Maddy, F. M. Chambers and R. P. Evershed, n-Alkane distributions in ombrotrophic mires as indicators of vegetation change related to climatic variation, *Org. Geochem.*, 2000, **31**(2–3), 231–235.

153. C. Dalton, H. J. B. Birks, S. J. Brooks, N. G. Cameron, R. P. Evershed, S. M. Peglar, J. A. Scott and R. Thompson, A multi-proxy study of lake-development in response to catchment changes during the Holocene at Lochnagar, north-east Scotland, *Palaeogeog. Palaeoclimatol. Palaeoecol.*, 2005, **221**(3–4), 175–201.

154. R. P. Philp, The emergence of stable isotopes in environmental and forensic geochemistry studies: a review, *Environ. Chem. Lett.*, 2007, **5**(2), 57–66.

155. R. P. Philp and T. Kuder, Biomarkers and stable isotopes in environmental forensic studies, in *Methods in Environmental Forensics*, ed. S. M. Mudge, Taylor and Francis, 2008.

156. R. H. Michener and D. M. Schell, Stable isotope ratios as tracers in marine aquatic food webs, in *Stable Isotopes in Ecology and Environmental Science*, ed. K. Lajtha and R. H. Michener, Blackwell Scientific, Oxford, 1994, pp. 138–157.

157. K. Lajtha and J. D. Marshall, Source of variation in the stable isotopic composition of plants, in *Stable Isotopes in Ecology and Environmental Science*, ed. K. Lajtha and R. H. Michener, Blackwell Scientific, Oxford, 1994, pp. 1–21.

158. R. R. Doucett, J. C. Marks, D. W. Blinn, M. Caron and B. A. Hungate, Measuring terrestrial subsidies to aquatic food webs using stable isotopes of hydrogen, *Ecology*, 2007, **88**(6), 1587–1592.

159. S. M. Mudge, Multivariate statistics in environmental forensics, *Environ. Forensics*, 2007, **8**, 155–163.

160. S. Wold, C. Albano, W. J. Dunn, U. Edlund, K. Esbensen, P. Geladi, S. Hellberg, E. Johansson, W. Lindberg and M. Sjöström, Multi-variate data analysis in chemistry, in *Chemometrics: Mathematics and Statistics in Chemistry*, ed. B. R. Kowalski, D. Reidel Publishing, Dordrecht, 1984.

161. M. B. Yunker, R. W. Macdonald, D. J. Veltkamp and W. J. Cretney, Terrestrial and marine biomarkers in a seasonally ice-covered Arctic estuary: integration of multivariate and biomarker approaches, *Mar. Chem.*, 1995, **49**(1), 1–50.

162. P. Geladi and B. R. Kowalski, Partial least squares regression: a tutorial, *Anal. Chim. Acta*, 1986, **185**, 1–17.

163. S. M. Mudge, Reassessment of the hydrocarbons in Prince William Sound and the Gulf of Alaska: identifying the source using partial least squares, *Environ. Sci. Technol.*, 2002, **36**(11), 2354–2360.

164. S. M. Mudge, G. F. Birch and C. Matthai, The effect of grain size and element concentration in identifying contaminant sources, *Environ. Forensics*, 2003, **4**(4), 305–312.

165. W. A. Burns, P. J. Mankiewicz, A. E. Bence, D. S. Page and K. R. Parker, A principal-component and least-squares method for allocating polycyclic aromatic hydrocarbons in sediment to multiple sources, *Environ. Toxicol. Chem.*, 1997, **16**(6), 1119–1131.

166. OECD SIDS Initial Assessment Report for SIAM 22. Long Chain Alcohols Category. OECD, Paris, 2006, 112,12 Annexes.

167. Exposure and risks screening methods for consumer product ingredients, Soap and Detergent Association, 2005, available from http://cleaning101. com/about/Exposure_and_Risk_Screening_Methods_for_Consumer_ Product_Ingredients.pdf.

168. Guidelines for exposure assessment, US Environmental Protection Agency, 1992, FRL-4129-5.

169. G. Veenstra, C. Webb, H. Sanderson, S. E. Belanger, P. Fisk, A. Nielsen, Y. Kasai, A. Willing, S. Dyer, K. Stanton and R. Sedlak, Human health risk assessment of long chained aliphatic alcohols, *Ecotoxicol. Environ. Safety*, 2008, in press.

170. Y. Iwata, Y. Moriya and T. Kobayashi, Percutaneous absorption of aliphatic compounds, *Cosmet. Toiletries*, 1987, **102**(2), 53–68.

171. *In vitro* skin permeation of radiolabeled 1-tetradecanol (myristyl alcohol) from an oil-in-water emulsion, Procter and Gamble Company, 2007.

172. Survey of use and exposure information provided by the member companies of the Long Chain Aliphatic Alcohols Consortium. *HPV Task Force, Long Chain Aliphatic Alcohols Survey*, Soap and Detergent Association, Washington, DC, 2002.

173. C. E. Cowan, D. Mackay, T. C. J. Feijtel, D. Van de Meent, J. D. di Guardo and N. Mackay, *Multi-media Fate Models: A Vital Tool for Predicting the Fate of Chemicals*, SETAC Press, Pensacola, FL, 1995.

174. W. F. Holman, Estimating environmental concentrations of consumer product components, in *Aquatic Toxicology and Hazard Assessment, Fourth Conference*, American Society for Testing and Materials, Philadelphia, PA, 1981.

175. T. C. J. Feijtel, G. Boeije, M. Matthies, A. Young, G. Morris, C. Gandolfi, B. Hansen, K. Fox, M. Holt, V. Koch, R. Schroder, G. Cassani, D. Schowanek, J. Rosenblom and H. Niessen, Development of a geography-referenced regional exposure assessment tool for European rivers: GREAT-ER contribution to GREAT-ER #1, *Chemosphere*, 1997, **34**, 2351–2374.

176. X. Wang, M. Homer, S. D. Dyer, C. White-Hull and C. Du, A river water quality model integrated with a web-based geographic information system, *J. Environ. Manag.*, 2005, **75**, 219–228.

177. T. W. Federle and N. R. Itrich, Comprehensive approach for assessing the kinetics of primary and ultimate biodegradation of chemicals in activated sludge: application to linear alkylbenzene sulfonate, *Environ. Sci. Technol.*, 1997, **31**(4), 1178–1184.

178. C. Schäfers, U. Boshof, H. Jürling, S. E. Belanger, H. Sanderson, S. D. Dyer, A. M. Nielsen, A. Willing, K. Gamon, Y. Kasai, C. V. Eadsforth, P. R. Fisk and A. E. Girling, Environmental properties of long-chain alcohols: 2. Structure–activity relationship for chronic aquatic toxicity, *Ecotoxicol. Environ. Safety*, 2008, in press.

179. S. E. Belanger, H. Sanderson, P. R. Fisk, C. Schäfers, S. M. Mudge, A. Willing, Y. Kasai, A. M. Nielsen, S. D. Dyer and R. Toy, Assessment of the environmental risk of long chain aliphatic alcohols, *Ecotoxicol. Environ. Safety*, 2008, in press.

180. S. D. Dyer, J. R. Lauth, S. W. Morrall, R. R. Herzog and D. S. Cherry, Development of a chronic toxicity structure activity relationship for alkyl sulfates, *Environ. Toxicol. Water Quality*, 1997, **12**, 295–303.

181. S. D. Dyer, D. T. Stanton, J. R. Lauth and D. S. Cherry, Structure–activity relationships for acute and chronic toxicity of alcohol ether sulfates, *Environ. Toxicol. Chem.*, 2000, **19**, 608–616.

182. S. E. Belanger, P. B. Dorn, R. Toy, G. Boeije, S. J. Marshall, T. Wind, R. Van Compernolle and D. Zeller, Aquatic risk assessment of alcohol ethoxylates in North America and Europe, *Ecotoxicol. Environ. Safety*, 2006, **64**, 85–99.

183. S. D. Dyer, H. Sanderson, S. W. Waite, R. Van Compernolle, B. Price, A. M. Nielsen, A. Evans, A. J. Decarvalho, D. J. Hooton and A. J. Sherren, Assessment of alcohol ethoxylate surfactants and fatty alcohols mixtures in river sediments and prospective risk assessment, *Environ. Monit. Assess.*, 2006, **120**, 45–63.

Subject Index

Note: page numbers in *italic* refer to tables or figures.